Emotional Ignorance

Dean Burnett is a neuroscientist, blogger, sometimes-comedian and author. He lives in Cardiff, and is currently an honorary research associate at the Cardiff University Psychology School. His previous books, *The Idiot Brain* and *The Happy Brain*, were international bestsellers published in over twenty-five countries. His *Guardian* articles have been read over sixteen million times and he currently writes the 'Brain Yapping' blog for the Cosmic Shambles network.

Dr Dean Burnett

Emotional Ignorance

Lost and Found in the Science of Emotion

First published in the UK in 2023
by Faber & Faber Ltd
Bloomsbury House
74–77 Great Russell Street,
London WC1B 3DA

Typeset by Typo•glyphix, Burton-on-Trent DE14 3HE
Printed and bound by CPI Group (UK) Ltd, Croydon CR0 4YY

A CIP record for this book
is available from the British Library

ISBN 978–1–783–35173–2

Printed and bound in the UK on FSC paper in line with our continuing
commitment to ethical business practices, sustainability and the environment.
For further information see faber.co.uk/environmental-policy

2 4 6 8 10 9 7 5 3 1

For Peter William Burnett
Love you, Dad

Contents

Introduction

This is not a book I planned to write.

In a sense, it's not a book I ever *wanted* to write.

But I'm very glad I wrote it. Doing so may have been the best thing I've ever done.

Confused? I don't blame you. Because so was I. That's how this whole thing started.

Let me back up and provide some context.

This is a book about emotions. Originally, it was about emotions in general, the science behind them, and how they work in the brain. It was going to be called *Emotional Intelligence*, because that's a common phrase, but it was also all about the science of emotions, which is smart stuff. Clever, eh?

However, there was a problem. I had assumed, apparently like many scientists and self-described intellectual types, that, scientifically speaking, emotions aren't really that complicated. Not like thoughts or memory or language or senses, the 'important' stuff that happens in our brain. They're a holdover, or a hindrance, if anything. Therefore, writing a book explaining them shouldn't be too tricky.

However, these assumptions were quickly proven to be substantially, hilariously, wrong. As soon as I started my research, I found that for every study that supported what I thought was an established fact about emotions, there were usually five more which said it wasn't. And all for different reasons.

I eventually had to face a hugely inconvenient, but irrefutable, fact: my knowledge about emotions was woefully insufficient for writing a book about them. Unfortunately, I was still contractually obliged to do exactly that. It was a tricky situation.

But then, in 2020, the COVID-19 pandemic happened, and the world went into lockdown as the virus tore across the globe. At first, I felt I was well placed to ride things out. I already worked from home, my job wasn't under threat, my wife and children and I are quite a harmonious group. This will be fine, I figured. This will be fine.

Then, in March that year, my father contracted the virus. Eventually, he was admitted to hospital. And I couldn't do anything. I couldn't help, I couldn't go to see him. It was a pandemic; we were all locked down, hospitals were quarantined, and all medical staff were working desperately hard to save lives.

Meanwhile, I was stuck at home, being updated about my father's condition via second- or third-hand messages, or over the occasional succinct phone call. But mostly, I was essentially trapped. It was just me . . . and my emotions. Emotions I was unfamiliar with, that I didn't know how to handle. Sure, I'd felt worry, concern, fear, and anxiety many times. But not like *this*.

And then, my fifty-eight-year-old father, with no prior health conditions, died. I never got to see him, or say a proper goodbye. And I had to endure the fallout of that – the worst emotional pain and trauma of my life – alone. Cut off from the world, and any possible source of help or reassurance. It was, to put it mildly, hell.

What happened then was, in the midst of the most powerful grief and emotional pain I'd ever encountered, my neuroscientist training kicked in. The geeky, relentlessly rational part of my brain somehow made itself heard amidst everything else going on in my head, and made the following compelling argument.

I am an experienced neuroscientist and science communicator, who currently has a brain full of powerful, nigh on overwhelming emotions, and who also has to *write a book about emotions*! Logically, I should take advantage of this incredibly unlikely combination of factors, and put it to use. Study the stress, pain, and uncertainty I was feeling, look at what it was doing to me, then try to explain why all this happens, what it means, and what the implications of it all could be. I could put my own feelings under the microscope – in the name of science.

And that's what I did. It turned out to be quite a journey. Exploring my own grief and unpicking why I was going through what I was going through took me to some very unlikely places. It also raised a plethora of intriguing questions.

Why do we humans look the way we do?

Why do our brains see what they see?

How come music affects us the way it does?

What propels a lot of scientific discovery?

How come our modern world is plagued by misinformation and 'fake news'?

Emotions, it turns out, are the answer to all of the above – and much more besides. My investigations into this all-pervasive aspect of our inner lives took me to the dawn of time, and the end of the universe. To the boundaries of fantasy and reality. From the most basic processes of life to the cutting edge of technology, and everything in between.

Because, as it turns out, far from being irrelevant, or of peripheral importance, emotions are a vital part of everything we are, and everything we do. They've shaped us, they guide us, they influence us, they motivate us, and, yes, they confuse us.

I had no idea about any of this when I started. I had no right to describe myself as emotionally intelligent. I was actually very emotionally ignorant. That's why I wrote this book. It's part scientific exploration, part grief journal, part 'journey of self-discovery', and more.

I'm not exaggerating when I say that writing this book kept me back from the brink during the worst time of my life, by helping me tackle my own emotional ignorance. And that's why the book's called that. If I can help you reduce yours too, even slightly, without you having to go through what I went through, then I'll consider it a job well done.

Dean

1

The Emotional Basics

When I first sat down to write this book with a head full of grief, my ultimate goal was to understand the emotions I was feeling, why were they happening, what were they doing to me, how they came about, and more. That, admittedly, is a pretty big ask.

Where do you even begin when you want to find out about how emotions really work? Well, if my previous scientific experience was anything to go by, you explore the fundamentals – 'the basics', if you like – and you build up a more complex and thorough understanding from there.

And when it comes to emotions, the most basic question of all is 'What is an emotion?' You can't do anything if you've not provided an answer to that question, right? So, that's the first thing I did.

Or at least, that's the first thing I *tried* to do.

I quickly encountered a problem, though: amazingly, and despite centuries of study and debate, there doesn't yet seem to be any robust consensus on what an emotion actually *is*. Which makes studying them somewhat tricky, to say the least.

Given how fundamental emotions are for everyone, and that they're estimated to have existed in some form for over 600 million years, you'd think that we'd have them figured out by now. But then, we've also been having and raising children for as long as our species has existed. Therefore, at this point we should all know and agree on the best way to raise a child.

But go to any online discussion about breastfeeding, sleeping arrangements for babies, or anything like that, and it's regularly a virtual bloodbath, like two rival guerrilla armies that have stumbled upon each other in an abandoned warehouse, albeit with more mentions of 'formula milk'.

That's not to say there's zero agreement among the relevant experts, that we know *nothing* about emotions. We're more ignorant about them than you might expect, but not *that* ignorant. Even so, there's a lot more emotional ignorance out there than you'd expect.

To understand more about why we still seem no closer to consensus on such a fundamental point, my first port of call was Dr Richard Firth-Godbehere, professional historian of emotions and author of *A Human History of Emotion*.[1]

I mentioned my difficulty finding an agreed definition for the subject of his life's work, at which Dr Firth-Godbehere laughed bitterly, like a war veteran listening to someone brag about how intense things got at the company paintball tournament. Paraphrasing the prominent emotion researcher Professor Joseph LeDoux, he told me:

> There are as many different definitions of emotions as there are people researching emotions. Possibly more, as people keep changing their minds.

I've been involved in the world of academia and science for most of my adult life, so I know that professional scientists and academics constantly disagree.* It's their favourite pastime, after consuming free wine at a conference reception.

But even so, I reasoned there must be *some* consensus in the field of emotion research, right? Neuroscience wouldn't function at all if

---

* Hence my laugh of recognition when Dr Firth-Godbehere shared with me the joke: 'What do you get if you put two historians in a room? Three opinions.'

nobody could agree on which organ the brain is, with some of us sure it was that wrinkly thing in the skull, while others insisted it was those long wriggly tubes in the abdomen. The whole discipline would be chaotic nonsense, and nothing would get done.

Nonetheless, while not *that* bad, the field of emotion research is indeed riddled with such uncertainties. Nobody denies emotions exist, but our understanding and concept of emotion is constantly changing and evolving over time, in ways that can be quite surprising.

And this is far from a new problem. I'd read so many modern reports about scientists and psychologists 'turning their attention to emotions' in 'recent decades', that I'd assumed emotion research was about 100–150 years old.

In truth, however, the study of emotions goes back *thousands* of years. Dr Firth-Godbehere identifies its starting point as having been with the Stoics, followers of Stoicism, one of the many philosophical schools of thought produced by the Ancient Greeks.

Founded in the third century BC by Zeno of Citium, the main thrust of Stoicism was accepting the natural state of things, living in the moment, and applying logic and reason in all circumstances.[2]

What with their enthusiasm for constant reason and logic, the Stoics also spent a lot of time pondering and studying emotions,[3] insofar as they could with the facilities and approaches available at the time.* They were among the first to recognise emotions as separate 'things', aspects of the human mind *distinct* from thinking and behaviour.

Predictably, Stoics often regarded emotions as unhelpful, identifying particular 'passions' including lust, fear, distress, and

---

* As advanced as they were, the Ancient Greeks didn't have brain-scanning technology.

delight, and declaring them to be irrational, contrary to Stoic ideals.[4] Such passions were to be resisted, because they cause people to perceive and behave towards things as they *want* them to be, not as they *are*.

This is a reasonable conclusion. For example, someone in the grip of lust can be rejected by the object of their affections multiple times, yet still pursue them nonetheless, because they want the situation to be different to how it really is, to what their eyes and ears are repeatedly telling them it is. Such behaviour is irrational, so against Stoic teachings (and also, often, the law).

Stoics felt passions led to *pathos*, an affliction one suffers due to excessive passions interfering with the ability to reason.[5] The only way to avoid pathos was to control or suppress the passions. They also believed that the way to truly avoid suffering was *apatheia*, the ultimate goal of Stoicism, a state of clear-mindedness where you're able to think and react logically and reasonably, in all situations.[6] Basically, Stoics were prototype Vulcans, two millennia before *Star Trek*.

Sadly, Ancient Greek civilisation eventually came to an end, taking the Stoics with it. However, they had quite a legacy, and their impact is still visible today. Important aspects of modern cognitive behavioural therapy stem from the teachings of Stoicism.[7] In the English language, we still use the word 'stoic' to describe someone unflappable, and 'pathos' to describe a quality that stirs up feelings of sadness or sorrow. *Apatheia*, in turn, is the distant ancestor of 'apathy'. It's a bit of a decline, admittedly, to go from meaning 'the ultimate expression of human consciousness' to 'can't be bothered'. Time is a great leveller.

Why, though? Why did one particular strand of Ancient Greek philosophy end up having such an impact on modern society? Well, Stoic principles endured largely because they were widely integrated into religion, particularly early Christianity.[8] For instance, Stoics, no fans of irrational lust, believed sex was only for reproduction during marriage.[9] Much of Christianity still agrees. There are also many parallels between Stoicism and Buddhism, which focuses on achieving enlightenment by extinguishing all earthly desires through mental discipline and meditation.

Although, Buddhism was founded by Siddhartha Gautama some 300 years *before* the introduction of Stoicism. So why not credit Buddhists with originating the study of emotions?

It's a fair question, and there may be some cultural bias at work here, but one thing the Stoics had going for them was a *materialistic* worldview: they believed that only things with a physical presence can be said to truly 'exist'. And because when we experience emotions our heart rate increases, we cry, we blush, we smile, etc., the Stoics believed that emotions had a *physical* presence. That means it's theoretically possible to identify and study emotions objectively. *Scientifically.*

Religion doesn't work that way. Buddhism, for all its positives, still includes concepts like karma and reincarnation, and, whatever your thoughts about such things, it's hard to reconcile belief in the intangible, or spiritual, with objective analysis and hard data. Unfortunately, the co-opting of Stoicism into more (Western) religious principles and worldviews also meant more of the former, and less of the latter.

Essentially, religion maintained and even furthered interest in emotions over the centuries after the decline of Stoicism. But this meant emotions regularly got tangled up with theological and faith-based priorities and practices. Not great for scientific understanding.

However, emotions weren't called that back then. They were 'passions', 'sins', 'appetites', 'drives', and so on. This remained the case until, in the nineteenth century, scientists got involved and staked a claim on the subject, by declaring that all those things were now called 'emotions', the term we still use today (for better or worse).

This 'rebranding' was initiated via the popular lectures of Edinburgh Professor of Moral Philosophy and qualified medic Thomas Brown, regarded by some as 'the inventor of emotions'.[10] When books of Brown's lectures entered circulation in 1820, his approach – subsuming all the previous 'passions', 'appetites', and 'affections' under the single category of 'emotions' – caught on.

This was reinforced by another Scottish philosopher/scientist, Professor Alexander Bain, founder of *Mind*, the first journal of psychology and analytical philosophy. In his 1859 book *The Emotions and Will*,[11] which many consider to be the first book about the psychological science of emotions, he wrote:

Emotion is the name here used to comprehend all that is understood by feelings, states of feeling, pleasures, pains, passions, sentiments, affections.

This scientific takeover of emotions was boosted further by yet another contemporary Scottish philosopher/professor, Sir Charles Bell, the man Bell's palsy is named after.[12] His interest in the facial nerves and muscles led him to study the facial expressions caused by emotions, which helped solidify the view of emotions as tangible, physiological processes, rather than spiritual, metaphysical things.

Bell's work and subsequent discoveries led to another influential book, *The Expression of the Emotions in Man and Animals*, written by a certain Charles Darwin.[13]

All this helped establish emotions as something that had physical basis in the real world, and could therefore be studied. The Stoics adopted this stance thousands of years earlier, but it was the nineteenth-century Scottish scientists who really cemented this as an accepted 'fact', as Dr Firth-Godbehere explained:

What Thomas Brown did, he put emotions in the brain rather than the soul – made them physical brain things in a more concrete way than those before him.

You might think this would provide clarity for the scientific study of emotions. In many ways, it did. But also, it didn't.

Following this reclassifying of existing mental phenomena as emotions, in 1880 Reverend Doctor James McCosh (*another* prominent Scottish philosopher) published his own book *The Emotions*,[14] with over one hundred examples of feelings, urges, longings, reactions, etc. that fell into the newly established category of emotions.

That's a lot of things. But was there a thorough, understandable, and consistent definition of emotion that applies equally to all of them? Something that would allow you to accurately determine what the label should and shouldn't be applied to?

No. And there still isn't. Coming up with one has, thus far, been a considerable challenge for the scientists and experts concerned. Indeed, as Thomas Brown himself once said: 'The exact meaning of the term emotion, it is difficult to state in any form of words.'[15]

In 2010, psychologist Dr Carroll E. Izard[16] interviewed numerous different experts from various areas of emotion research, to find what (if any) consensus there was regarding the definition and properties of emotions. The eventual summary this study produced was as follows:

Emotion consists of neural circuits (that are at least partially dedicated), response systems, and a feeling state/process that motivates and organizes cognition and action. Emotion also provides information to the person experiencing it, and may include antecedent cognitive appraisals and ongoing cognition including an interpretation of its feeling state, expressions, or social-communicative signals, and may motivate approach or avoidant behaviour, exercise control/regulation of responses, and be social or relational in nature.

If you're anything like me, reading this left you *more* confused about what emotions actually are, not less. In fairness, it isn't meant to be a definition, rather a summary of what current experts agree are the consistent features of emotions. Even so, it gives an indication of why our understanding of emotions, particularly in the scientific context, is still so limited, even though the average person is very familiar with them and seems to understand them intuitively.

In essence, from a scientific perspective, the label 'emotions' is like the label 'farm animals'. We all know what a farm animal is; cows, horses, sheep, chickens: those are farm animals. Eagles, octopuses, crocodiles: they are *not* farm animals.

But scientists studying emotions are like vets responsible for treating a sick farm animal. They need to know specifics, or they can't do their job. You can't just say, 'The farm animal is ill'. Is it a cow? Chicken? Dog? Pig? These all need to be treated in very specific ways.

And because of the slippery, uncertain, often intangible nature of emotions, it's like the vets in this analogy can't even go to the farm and look for themselves; they must do it all over the phone.

Ironically, one thing emotion researchers *do* genuinely agree on is that a reliable definition of emotions, one that works for everyone, would be very useful indeed. But such a thing, for now at least, seems constantly out of reach.

The work continues, though. Emotion researchers are finding out more about how they work all the time, and presumably they'll be able to clarify exactly what they *are*, eventually.

One unexpected positive about the persistent confusion around emotions was that it gave me some perspective on my own emotional ignorance. I may not have had a clue about how they work, but apparently the same can be said of a lot of people. Even the experts. So, there's that. But still, it was a concern for me and my objectives.

However, this wasn't as big a problem as it may seem. It's actually familiar ground for neuroscientists like myself. After all, as with emotions, it's very difficult to specifically define things like thoughts, minds, sensations, and so on. Most of the important things our brains do are slippery and intangible in nature. But we still study those all the time.

How? By focussing on the tangible, on the things we *can* see, assess, measure, and define. In this case, we focus on the biological, physiological processes that occur when we experience emotions. We may not need a specific verbal definition of what emotions are, if we can see what's going in our brains and bodies when they occur. That will give us a much better idea of what they are, and do.

The philosophers and historians had done their part, but now it was time to let the scientists take over the exploration of emotions. Let's just say, I had a good feeling about this . . .

A body of emotions

When my father was taken into hospital, I didn't cry.

I *wanted* to. I was really worried about him, and beside myself with frustration about the dire situation we were all in. And this wasn't some macho posturing thing; at the time, I was stuck at home with my wife and two small children, so such posturing would have been a complete waste of time even if I were that way inclined.

But nonetheless, I didn't cry. Not right away at least. I did eventually, but in brief fits and starts. And if I'm being honest, when I did cry, contrary to what is often asserted, I didn't feel much better. I was as upset as before, but now with wet, red eyes and a leaking nose. I was also making weird noises that alarmed my neighbours. Overall, crying didn't really improve my situation.

I was dwelling on this because of what I'd read about the Stoics, and their thoughts on emotions. Specifically, their conclusion that emotions are distinct, tangible things, because they're expressed in consistent, specific ways by our bodies. We don't just experience emotions mentally; we express them physically, often without meaning to.

I figured that if I could find out *why* this happens, why emotions have such physical effects on our bodies, maybe that would help clarify what emotions actually are, how they work, and why they were affecting me so.

And one very familiar, and overt, example of emotions leading to reactions in the body, is crying.

So, why do we cry?

I don't mean 'what things make us cry?', because that applies to everything from chopping onions to dust in the air, from a heart-breaking loss to receiving a swift kick in the gonads. No, I mean, why do we cry *at all*? Why did evolution think leaking water from the eyeballs was a useful ability?

Here's the thing: for all that it's a common, fundamental thing we all do, even if we just take crying to mean 'producing tears',* it's surprisingly complex.

For instance, humans have *three types* of tears.[17] There are basal tears, the fluid produced constantly that forms the three-micrometre-thin liquid film coating our eyes at all times, keeping them clear, lubricated, and healthy.[18, 19]

When dust, grit, or vapour from chopped onions gets into our eyes, we produce reflex tears, to clear the ocular intrusion, like using the shower to flush a spider down the plughole.

---

* And ignore all the snot and weird noises that many emit when crying.

Finally, there are psycho-emotional tears, produced when we experience powerful emotions: usually sadness, but also anger, happiness, and others. But while the other tear types have obvious functions, what's the purpose of tears when we're sad? You can't wash a negative emotion out via your eyes (or so I'd always assumed).

There are many theories as to what function psycho-emotional tears serve.[20] One is that they broadcast our emotional state; they display to those around us that we need help. Or, that we're available to help, or share, if it's a positive emotion.

But then, research reveals that tears caused by emotions are chemically different to those produced via eye irritation.[21] If tears were purely for display purposes, just there to be looked at, this wouldn't be necessary.

Emotional tears contain oxytocin and endorphins, 'feel good' chemicals, that improve mood when absorbed through the skin.[22] That's presumably handy when you're sad. However, producing very small doses of such chemicals and dribbling them down our cheeks is a rather inefficient means of administration. It's presumably impossible to get high off your own tears (although that would explain the popularity of misery memoirs).

Other studies show that, when inhaled, women's tears suppress arousal and testosterone levels in men.[23] It's unclear if the same thing happens with women and male tears, but it's not unheard of for women to show behavioural changes[24] after inhaling other people's secretions.* Either way, it does suggest that our emotional tears are *chemically influencing* those around us. Which is somewhat creepy.

This also shows that the connection between our emotions and our physiology is much deeper and more profound than many may assume. Far from being just abstract, intangible products of our

---

* Whatever you're thinking of, it's nowhere near that vulgar.

minds, with no more physical substance than our shadows, our emotions can affect our bodies at the most fundamental biochemical levels.

Obviously, I'm not the first to notice this. As we saw, the Stoics were flagging this up millennia ago. And it's also evident in how much the language around emotion centres on organs and body parts that *aren't* the brain.

All things romantic refer to the heart, and conversely we experience 'heartache' or 'heartbreak' when romance goes wrong. We have 'gut' feelings, for decisions or inclinations arrived at instinctively, unthinkingly, often via emotions. Powerful emotions can leave us 'breathless', bringing the respiratory system into play. Angry ranting is often referred to as 'venting your spleen'. Happiness regularly brings about a state of calm and relaxation, suggesting a drop in muscle tension. Or, if we're highly amused, it can lead to 'belly laughs'. And how many ways of describing fear are just variations on 'I soiled myself'? Our bowels and waste systems respond to emotions too, even though we'd rather they didn't.

It's often a useful shorthand for people, particularly neurobods like myself, to differentiate between brain and body as if they're separate things, with the brain piloting the body like it's some elaborate meat vehicle. But the implication that they're completely distinct entities is wrong; as the many links between emotions and bodily functions reveals, they're extensively intertwined and overlapping.

After all, the brain, for all its powers, is still an organ. It needs the body in order to survive and function. The upshot of this is, while the brain undeniably controls and influences the body, the reverse is also often true, and our body influences our brain in various ways.

The central nervous system, the brain and spinal cord, is located within the skull and spine, hence damage to these areas can be so significant (and devastating). But the central nervous system interacts with the rest of the body via the peripheral nervous system,[25] another complex network of nerves and neurons, which links the central nervous system to all the other organs and tissues.

It has two components. One is the somatic system, which conveys sensory information from the organs (temperature, pain, pressure, etc.), and sends motor signals to the muscles, allowing us to consciously move our bodies.[26]

The other is the autonomic nervous system,[27] which oversees unconscious processes: anything that happens without us thinking about it, like sweating, heart regulation, liver function, etc.

The autonomic nervous system is itself made up of two distinct parts, the sympathetic and parasympathetic nervous systems. The sympathetic system fires our internals up to deal with dangers and threats; it induces the famous 'fight or flight'* response.[28] The parasympathetic system essentially does the opposite: it keeps our biological processes in a calm, relaxed, 'baseline' state, often termed 'rest and digest'.[29] The general activity within our bodies and organs is maintained by a careful balance between these two autonomic systems.

Here's the cool bit: these peripheral nervous systems are largely (although not *entirely*) regulated by the brain. Specifically, by one of the deeper, more fundamental brain regions, the hypothalamus,[30] a key region responsible for 'controlling' what's going on in the body.

Accordingly, the hypothalamus also oversees the endocrine system,[31] which is where our brain influences our metabolism and bodily function via hormones: chemicals secreted into the bloodstream. In a sense, the endocrine system is to the nervous system what physical or 'snail' mail is to email. As in, they both send and receive information, they just differ in terms of speed and capacity.

* Although now it's fight, flight, or *freeze*. Many species opt to remain completely motionless when faced with a threat, which is often just as useful a reaction.

The point of telling you this is that these unconscious auto-nomic and endocrine systems are how our emotions influence what's going on our body.[32] That's why an emotional experience invariably comes with many physical aspects: your heart rate alters; your stomach clenches or you feel sick; you cry; your skin flushes or goes pale as blood is directed to it or away from it; and you feel the urge to, shall we say, 'relieve' yourself. These are all functions of the autonomic nervous system, the activity of which is often beholden to a brain, a brain that's often experiencing potent emotions.

But it's not all one way. Weird as it may sound, our bodies can also influence the emotions occurring in our brain. The tail does indeed wag the dog, and surprisingly often.

Obviously, if you break your toe, get food poisoning, or a vicious cold, this makes you very miserable, or angry. That's technically your body dictating the emotions your brain is producing. However, I'm referring to more complex, subtle, and less indirect ways in which your body can influence your emotions.

For instance, you may have heard the term 'hangry', the phenom-enon where you're more irritable or grumpy when you're hungry. It may seem like this is just some social media-friendly portmanteau, but being hangry is a legitimate phenomenon, according to numerous studies. One eyebrow-raising experiment, led by Professor Bushman of Ohio State University, found that married couples show greater levels of aggression towards each other* when they have lower blood sugar.[33]

It makes sense; the brain depends on glucose, i.e. blood sugar, to do everything it does, and things go awry when it can't get enough.[34] So, logically, blood glucose levels in the body affect what the brain can do, like exert self-restraint and control aggressive

---

* This particular study assessed this by recording how many pins were stuck into voodoo dolls representing a subject's partner. Last I checked, this type of measurement wasn't included in the metric system.

impulses. And blood glucose levels are determined by the digestive system, as well as the liver and muscles, and the myriad hormones they secrete and respond to.[35] So, there's one instance of the other organs dictating what the brain can do, emotionally.

In fact, the digestive system is getting much attention lately for its surprisingly important role in our mental state and emotions. Rather than just a long wobbly tube that food passes through, the digestive system is incredibly sophisticated, with a suite of specific hormones,[36] a dedicated branch of the nervous system (the enteric nervous system, which is so complex it's often dubbed 'the second brain'[37]), and *trillions*[38] of diverse bacteria forming the gut microbiome. Seriously, the digestive system could challenge the brain for the title of 'most influential organ'.*

Given all that, it's unsurprising that the digestive system seemingly wields considerable influence over brain function and mental health, thanks to what scientists have termed the 'gut–brain axis',[39] something providing a new frontier in health and wellbeing research, offering new avenues for treatments for conditions such as depression.[40] Far from being a meaningless cliché, it seems that *science itself* is 'following your gut'. That phrase is blatantly more valid than many assume. In any case, that's another way the body influences emotions.

But how? How, or even why, would the digestive system, or any other organ, so profoundly influence the emotional processes in the brain?

As well as the blood sugar factor, what goes on in the gut affects the chemical makeup of the whole body – it's where all the important chemicals we need to live enter the body, after all – and this would predictably have a knock-on effect on the brain, which responds and reacts to the chemical environment around it, like any other organ.

---

* Admittedly, it would lose that challenge. But still!

There's also our friend, the endocrine system. As well as producing and releasing them, our organic brain *responds* to hormones. And the gut, kidneys, liver, body fat, and more: they all produce hormones that our brain can detect and react to, meaning a direct effect on our brain and our emotions.*[41]

But there's an even more direct way for the body to influence the brain: via the vagus nerve.[42] This is one of the twelve cranial nerves, incredibly important nerves that emerge directly from the brain and connect it to important parts of the body, like the ears and eyes. They're vital conduits that relay crucial signals, like much of our sensory information, to the brain.

The vagus is the largest of the cranial nerves, because while most of the twelve connect to parts of the head and neck, the vagus connects the brain directly to nearly *all* organs lower down. It's the biggest part of the parasympathetic system, exerting direct influence over organs and tissues including the heart, lungs, digestive system, bladder, sweat glands, and more. And the reason it is so relevant to the physiology of emotion is that around 80–90 per cent of the neurons and fibres that make up the vagus nerve are *afferent*, meaning they carry information *from* the organ and *to* the brain.†

This means that there is a direct line of communication open between the lower organs (as in, those below the head) and the brain at all times. Essentially, the vagus nerve allows the brain to 'know' what's going on with all the different parts of the body, at any given moment, and to respond accordingly.

Have you ever wondered why some people say things like, 'My joints are aching, that means it's going to rain'? This could be why. Their joints may be responding to the drop in air pressure that

---

* If you still need convincing that hormones can influence emotions, speak to any teenager, pregnant woman, or vaguely aroused man.

† *Efferent* fibres do the opposite, carrying signals and commands from the brain to the organ/tissue.

comes before rain. This sensation is relayed to the brain via the vagus nerve, and the powerful brain recognises that this often happens right before a downpour, and puts it all together.

As you can imagine, vagus nerve activity, aka 'vagal tone', is a big factor in our emotions, particularly the physiological aspects.[43] It's believed to be the means via which the gut influences our mental health,[44] because if there's a problem in your vitally important gut, the vagus nerve means your brain knows about it immediately. And if the signals your brain is receiving from a very important source are just constantly saying, 'SOMETHING'S WRONG!', this would presumably cause a negative emotional reaction to occur, and to occur often.

Accordingly, vagus nerve stimulation is increasingly used as a treatment for depression and anxiety, both widespread disorders strongly linked to poor emotional control.[45, 46]

Here's a thing, though: even considering everything we've covered so far, if there's one thing that's consistent and everyone can agree on, it's that it's the brain that's generating emotions, right? The body may be sending important information which determines which emotions occur, but the brain's still the one producing them. Even if it is supplying the raw materials, the body isn't 'creating' the emotions, any more than the truck that delivered the bricks is responsible for the building of a house.

Because why would you expect any organ other than the brain to produce emotions? Where does that end? Are we going to make our lungs do maths? Or let our kidneys store memories? Or use our bladder to read a map?* Surely, if there's one thing that everyone researching emotions can agree on, it's that they come from the brain. However, even here the consensus isn't 100 per cent,

* Although given how often long car journeys are delayed by toilet stops, maybe this isn't quite so ridiculous.

because some scientists argue that the body is indeed responsible for 'creating' emotions.

There's a theory called the somatic marker hypothesis,[47] which argues that emotions come from the brain only after it receives *specific arrangements of signals from the body*. For example, something happens (e.g. we almost get mown down by a car while crossing the road), and, via information relayed from our senses, our body *reacts* (increases heart rate, tenses muscles, drains blood from face, etc.), often before our conscious brain processes have a chance to really 'think' about it, in any appreciable way.

These unconscious signals from the body, the heart rate and muscle tension and so on, are unavoidably relayed to our brain. These are the 'somatic markers'. Over time, the brain learns the particular emotional response that is required when the body produces these somatic markers. So, if we encounter something that causes our heart rate to go up and our muscles to tense again, that particular combination of somatic markers tells our brain to make us experience fear.

If anything, this suggests that emotions are determined more by the body than the brain. It's the specific assembly of bodily responses that dictates which emotion is experienced. The brain's job is to interpret them in a way that makes sense.

It's a subtle difference maybe, but an important one; going back to the previous 'building a house' analogy, it suggests the body's role isn't supplying the bricks to the builder of emotions that is the brain. Rather, it's the architect. The body supplies the blueprints of emotions, not the raw materials, and the brain follows the body's instructions for creating them.

It's intriguing, and there is some evidence for it,[48] but the somatic marker theory is by no means universally accepted. Many scientists have highlighted its limitations,[49] like how we regularly experience emotions *without* an event to trigger them.

We've all been there: you're strolling along, minding your own business, when suddenly, for no discernible reason, your brain dredges up the memory of a horrendously embarrassing experience from your past (usually from your teens), and you're left trying to cringe yourself inside out on the street corner. In such instances, there was nothing external for our body to react to. Yet, we often experience emotions without any obvious 'somatic markers' present. That surely undermines the eponymous theory, somewhat?

Proponents have addressed this problem by suggesting an 'as if' body loop,[50] where the brain effectively *simulates* somatic signals 'as if' the body is sending them, and can thus generate emotions independently. However, this is a rather inefficient process. Having to simulate what the body does before emotions can be induced adds several layers of 'admin' for the brain, an invariably frugal organ. This seems especially unlikely considering how immediate our emotional responses typically are.[51]

Overall, the somatic marker hypothesis is just one of the many theories out there regarding how emotions work, in the neurobiological sense. But that it's taken as seriously as it is shows that any ideas about the emotions being purely abstract processes, contained entirely within the mind and/or brain, need to be ditched.

Our emotions have a big impact on our physiology, and vice versa, from sadness changing the chemical composition of our tears to the bacteria in our guts being able to influence our mood. It's impossible to deny that our bodies are riddled with emotion, and that emotions clearly do have a tangible, physical presence. And if that's the case, it may be something that science can observe, record, maybe even control.

And this made me wonder: maybe that's what was stopping me from crying during such an emotionally fraught time? Maybe it wasn't my brain (which has always served me well, admittedly) that was 'out of whack', but my *body*. After all, it was the more physical aspect of my emotional reaction that I felt was absent, or insufficient. And I admit that while I've spent many years using my brain, I have kind of taken my body for granted.

I did start going to the gym more once I began working from home but, if anything, my body liked that even less. It certainly complained a lot. So maybe it's turned against me? Maybe it's gone on strike, so is denying me the critical emotional responses when I need them the most?

However, this explanation assumes there's a clear separation between my body and brain. And I've already specifically stated, repeatedly, that this is *not the case*! The whole point of me doing this was to reduce my emotional ignorance, not enhance it.

Also, if I keep anthropomorphising my own body as if it's some distinct entity rather than, you know, *me*, they're going to take my science licence off me.

Nonetheless, at this point it was impossible to deny that emotions have a far more 'physical' presence in our bodies than I'd ever appreciated, one that extends far beyond the boundaries of our brains.

And if that's the case, shouldn't it be possible, like it is with memories and sensations like pain, to at least get a rough idea of the fundamental physiological form specific emotions take within us? And to use that in turn, to figure out how they work, and how and why they affect us like they do?

A logical argument, no doubt. Unfortunately, I soon discovered that following through with it presented a rather considerable challenge. A challenge I would have to face up to.

Emotional face-off

When people heard the news about my father's hospitalisation, many got in touch to see how I was doing. This led to a lot of people asking me, 'How are you feeling?' And I gave them a strictly honest answer, which was, 'I don't know'.

Technically, this was accurate in two ways. I genuinely didn't know how to describe how I was feeling; I was in uncharted emotional waters for me, and I lacked the experience or vocabulary to convey it. But I also don't know how I feel things *in general*. As in, I don't know how feelings, emotions, work in the brain, and how we end up experiencing them. I was subtly admitting my own emotional ignorance.

For the record, I'm fully aware that nobody was actually asking me the mechanism via which I was experiencing emotions. But, in my defence, I was mentally in a very bad place, and if I want to use harmless but woefully analytical wordplay as a coping mechanism, then that's what I'll do!

It did make me wonder, though: what *should* I be feeling at that point? What is the correct and appropriate emotional reaction in this scenario? I should be sad, obviously. Or maybe scared? Or even angry, at the unfairness of everything? Or all three?

Can you combine these distinct emotions, and feel them all at once? Or do emotions have a 'one in, one out' policy? Is there a specific emotional reaction for every feasible scenario? Or do we have a sort of 'basic range' of emotions that we combine in interesting ways, like how the limited range of notes produced by piano keys can create many different concertos?

This question, it turns out, is a particularly important, and rather contentious, one in the field of emotion research.

Dr Tim Lomas's 'Positive Lexicography'[52] is an ongoing project to catalogue non-English terms for specific emotional experiences with no direct translation. The German *Schadenfreude* ('pleasure at the misfortune of others') is probably the most famous example of such a thing. Others include Norwegian *utepils* ('to sit outside on a sunny day and enjoy a beer'), Indonesian *jayus* ('an unfunny joke told so poorly that one cannot help but laugh'), and, from the language of my own country, Welsh *hiraeth* ('a particular type of longing for the homeland, or the romanticised past').

At present, the lexicon has *over a thousand* entries. Does that mean there are over a thousand distinct human emotions that humans experience?

Unlikely. They're arguably all variations on/combinations of more familiar, 'basic' emotions, given a unique label by a particular culture; for example, *utepils* is surely just a particular expression of happiness. We English speakers do this too; Dr Firth-Godbehere described the mix of fear and disgust we in the West have labelled 'horror'.

But if these thousands of emotional experiences are all combinations or variations of more fundamental ones, what *are* the fundamental ones? How far down can you go, before you hit emotional bedrock?

At present, nobody is completely sure. Yet this is likely to be a crucial point. When we discovered that germs were the basis of much illness and disease, it totally revolutionised medicine and public health, saving literally millions of lives. Maybe establishing the basic elements of emotions would yield similar gains, albeit of a more psychological slant, revolutionising mental health rather than physical health?

The emotion research community is seemingly split into two sides over this question. One side believes there are indeed a small number of basic emotions, innate to every human brain, which give rise to all the other known emotional states. The other side argues that there are essentially *no* basic emotions, that the fundamental substance of emotion is something deeper and more general called 'affect', and our brains learn to create emotions essentially 'on the fly', as and when needed.

Both have good reason to think what they think, and interestingly, a lot of the debate stems from a surprising source: the human face.

Our faces are important to us humans. That's a fact. Our brains have a dedicated neurological region, the fusiform face area,[53] specifically for recognising and reading faces. This helps explain why we're so good at recognising whether or not a smile is 'genuine',[54] or why eye contact is a vital element of trust and communication,[55] or why we see faces even when they're not there,[56] and so on. Our brains have evolved to utilise faces in so many situations and are constantly seeking them out.

Another crucial property of our faces? Displaying our emotional state. Our faces are constantly reconfiguring to produce expressions that reflect the emotion we're experiencing. That's why if someone's sad, angry, happy, disgusted, etc., we can usually tell just by looking at them.

This normally happens automatically, without us thinking about it. If anything, it's actually quite hard to consciously adopt a convincing facial expression for an emotion we're *not* experiencing. If you've been made to 'smile' for your 743rd successive wedding photo, you'll know this is true.

Because it happens consistently and involuntarily, it suggests a direct neurological link between our brains and our faces, allowing the emotions occurring in the former to be reflected in the latter (as noted by Charles Bell and Darwin in the nineteenth century[57]).

Therefore, logic suggests you can work out what's going on emotionally in the brain by studying the face,[58] like how you can learn a lot about an animal by the tracks it leaves through the undergrowth. Much of the most prominent emotional research rests on this premise.

The most influential scientist in this area is Dr Paul Ekman. Before his work in the 1970s, it was believed that facial expressions signifying emotions were learned from those around us,[59] much like how we acquire the words and language we're eventually fluent in. We're not born with them, they're not innate – it's nurture, rather than nature.

However, Ekman's studies showed that people from very different cultures often use the same facial expressions to display the same emotions.[60] This was important, because if facial expressions really were learned, cultural things, this would be like all the world's different cultures ending up speaking English, independently of each other. This is ludicrously unlikely,* a premise best kept to old *Star Trek* episodes.

Ekman's findings suggested that a far more likely explanation is that emotional facial expressions are a fundamental evolved property of the brain. Just like how the overwhelming majority of humans, regardless of background, end up with five fingers on each hand, we all have the same facial expressions for certain emotions. Nobody *learns* to grow five fingers.

Specifically, Ekman identified six emotions with the same facial expressions across cultures: Happiness, Sadness, Anger, Fear, Disgust, and Surprise. These were dubbed the 'basic' emotions, and are still regularly referred to as such today.

Initially, critics argued that many cultures having common facial expressions could instead be explained by the level of cultural cross-pollination which had occurred throughout human history, most of which had taken place long before Ekman's research in the 1970s.

In response, Ekman applied his research methods to the Fore people of Papua New Guinea, a remote tribal community that had experienced little contact with the outside world.[61] If Ekman's critics were right, and the reason most cultures used the same facial expressions was because they'd all learned them from each other over centuries of interaction, then the Fore people should have noticeably different expressions to everyone else. Because they'd experienced little to no cultural mixing, they would have their own unique emotional expressions.

---

* Although it's a scenario that persists in the minds of many British tourists.

And what do you know: the Fore people *did* use familiar facial expressions for specific emotions. In the realm of emotion research this put the theory of universal basic emotions front and centre. The six basic emotions theory has influenced and defined a great deal of research and development since then, in areas as diverse as psychological evaluation, facial recognition software, even marketing algorithms.

However, the six basic emotions theory is by no means without issues. For instance, why is 'surprise' included? It's more fleeting than most emotions, and linked to even more fundamental processes, like the startle response.[62] There's debate over whether surprise counts as an emotion at all, let alone a 'basic' one.[63]

This dispute isn't great for the basic emotion theory's credibility. It's like if someone claiming to be an expert in the history of popular music kept insisting that Homer Simpson was a founding member of the Beatles. That would cast doubt on everything else they said.

Similarly, a 2014 study from the University of Glasgow, using advanced computer modelling of expressions, reported that expressions of anger and disgust, and fear and surprise, have features in common, and therefore should be merged into one core experience, suggesting there are only *four* basic emotions.[64] These are just some of the challenging findings that have come to light.

Another issue is that while our faces undeniably display emotions, it doesn't automatically follow that all basic emotions cause involuntary facial expressions. What expression does a person experiencing pride, or satisfaction, have? Your face is also capable of adopting an expression you're *not* feeling, hence the term 'resting bitch face'.

Ekman himself has acknowledged this, later expanding his own system of basic emotions to include 'invisible' ones, like pride, guilt, embarrassment, etc.[65]

So, even among those who *support* the theory of basic emotions, there's uncertainty, disagreement, and dispute. But then, there are those who are unconvinced by Ekman's original findings and subsequent claims, due to issues and potential problems that have come up since.

For instance, the photos of facial expressions used in Ekman's research were of (American) actors who'd been told to look 'scared' or 'disgusted'. But is that a valid representation of how facial expressions of emotions usually work? Because when most people feel scared or disgusted, they don't put conscious effort into showing it on their faces, as mentioned earlier.

When similar studies were conducted which used candid shots (where people with emotional facial expressions were photographed discreetly), general recognition of what emotion was being displayed dropped from around 80 per cent right down to 26 per cent![66] Also, studies using more advanced modern methods have revealed that the facial expressions of different cultures, how they recognise and respond to them, *do* have some marked differences after all.[67]

The ramifications and interpretations of these studies can be discussed at length, but it looks increasingly like the idea of universal basic emotions, expressed and recognised via the face, is not the whole story. And there's a growing effort in emotion research to challenge its dominance.

One person spearheading these efforts is Professor Lisa Feldman Barrett of Northeastern University. In her book, *How Emotions Are Made: The Secret Life of the Brain*,[68] she explains how, as an aspiring researcher in the 1990s, she studied the effects of emotions on self-perception. Only, none of her experiments and studies were working, as subjects repeatedly failed to differentiate between sadness and fear, anxiety and depression.

According to the accepted wisdom, this shouldn't happen. Sadness and fear are basic emotions with universal facial expressions. Your average person should be able to tell them apart easily. And yet, every time Barrett tried to get her subjects to do that, they struggled. She eventually found an increasing number of other experiments and data reporting similar issues. It was then discovered that even minor changes to the methods used in Ekman's original ground-breaking experiments produced very different results.[69]

For instance, the original studies asked subjects to match a facial expression to an emotional statement, e.g. 'This person has just won millions of dollars', which would be matched with the 'happy' expression. But if you just gave subjects a photo and asked, 'What emotion is this person showing?', the average performance accuracy dropped through the floor.

Either Barrett and dozens of other experienced researchers were all doing something profoundly wrong, or the theory of basic universal emotions was *itself* flawed.

As a result, a growing number of researchers now contend that basic emotions *don't exist*. Instead, they propose the 'constructed emotions theory'. This argues that emotions, even what we'd label 'basic' ones, are not hard-wired in the brain, but created *in the moment*, as and when needed, based on raw sensory data, memory and experience, body responses, and anything else the brain has access to (which is a lot).

Although it seemingly flies in the face of common sense, the idea we 'make up' our emotions moment by moment is an increasingly accepted position, with ever more evidence in favour of it.*

Think about it: do we pull the exact same facial expression every time we feel a certain emotion? Any decent actor would tell you we definitely don't. Do we all experience the same emotional reaction to the same things? No way. There are songs or foods or artworks or individuals out there that inspire tremendous joy and pleasure in

---

* In my experience, many things described as 'common sense' are neither common nor especially sensible.

some, visceral loathing in others, and anything in between.

Even within ourselves, we don't always have the same emotional reaction to the same thing; context is key. Seeing your romantic partner can fill you with extreme happiness a week into your relationship, or agonising sadness a week into your breakup.

If, as Ekman's theories argue, our emotions were hard-wired, with accompanying facial expressions, they should be much more consistent than is demonstrably the case. Hence the increasingly prominent argument that the brain creates our emotions anew depending on the situation and context. Even if there is a direct connection between the emotions in our brain and the expression on the face, it's presumably a single thread in an exceedingly complex tapestry.

Also, the idea that our brain is spontaneously creating our emotions moment by moment is by no means far-fetched. For instance, our vision starts as simple pulses of neuronal activity, relayed to the brain by the retinas in our eyes, which can only detect three different wavelengths of visible light.[70] Basically, our eyes can only 'see' three colours. And yet, from this meagre information, our brains are constantly constructing an ever-changing rich and detailed visual experience.

Our brains are also believed to do something similar with memories, that they're regularly 'rebuilt' from discrete elements stored in the cortex, as and when needed.[71] This would explain why our memories are so pliable, so prone to shifting and changing with time and context.

Essentially, if our brains are constantly creating, from basic components, both our memories and our vision, why *wouldn't* they do the same with emotions? That's basically what the theory of constructed emotions, the constructivist view, is arguing.

In truth, the 'Basic emotions versus Constructivism' debate is far from settled. Both have much supporting evidence and, given the slippery and poorly defined properties of emotions – not to mention the difficulty of getting reliable hard data from the workings of the brain – a conclusive answer one way or the other remains a long way off.

It did make me wonder about my own emotional ignorance and incapacity, though. My inability to cry, my difficulty in recognising what I was feeling: what was the cause of that? The basic emotion theory suggests something could have gone awry with the fundamental circuitry in my brain. But if the constructivist argument is correct, it could be that my brain just hasn't figured out how to create and deal with the 'appropriate' emotional response yet, as I'd not gone through such an experience before.

The former implies a physical problem in my grey matter, which is unsettling, while the latter suggests a deficit I could remedy with patience and familiarity. I'd be lying if I said I didn't find myself leaning towards the constructivist theory as a result. But then I remembered that that's not how science works. You can't just choose one argument because it strikes you as more 'pleasing'.

Although, preferring one theory of emotions over another because it's more emotionally reassuring is amusingly ironic. It also suggests that I'm not quite so emotionally stunted after all.

But still, that's no excuse to abandon my scientific principles. It's no good trading emotional ignorance for regular ignorance. So, instead, my quest for emotional intelligence led me to an obvious follow-up question: if the body is reflecting the emotions happening within the brain, where in the brain are these emotions actually coming from?

Emotions in the brain

Isolating and observing a specific bit of the brain, and confirming that it's performing a specific function, is a fiendishly difficult process at the best of times. When what you're looking for is something that still thwarts efforts to scientifically define it, it's harder again.

There's also another issue that confuses matters: widely held assumptions and ideas about how emotions work in the brain, which we know are scientifically invalid, but still refuse to die, like one of Tolkien's elves. Or a particularly irritating bluebottle.

The most common example is probably the whole 'left brain/right brain' claim. This contends that the left side of the brain is logical and analytical, while the right is creative, expressive, *emotional*. So, if you're a reserved, stoic sort, you use your left brain more. Meanwhile, if you're more extroverted, emotive, and artistic, you use your right. This claim pops up in many an inane quiz on social media, which purport to give you a questionable psychological analysis via a few banal multiple-choice questions, or just having you stare at a revolving shape for a bit.

Let's be clear: this left brain/right brain claim is wrong. Or is, at least, an obscenely simplified view of how the brain operates. However, in my efforts to debunk it as thoroughly (and snarkily) as possible, I found that there are actually a few underlying scientific truths to it. I'll confess, this annoyed me.

Firstly, just to confirm, the human brain does have two sides, two hemispheres. That's why it resembles a pair of large walnuts glued together at the base. Or a set of mummified buttocks. Point is, there is indeed a distinct left and right side of our brains.

Exactly *why* the brain is like this is unclear, but for half a billion years pretty much all organisms have adhered to a symmetrical form, and there are numerous possibilities for what the advantage of this is.[72] But whatever the reason, our brains have distinct left and right hemispheres, connected via the corpus callosum, a thick band of white matter tracts that relays information between them, like a powerful (but squidgy) broadband cable.

There's evidence linking greater thickness of the corpus callosum to higher intelligence,[73] which makes sense: a thicker corpus callosum means more connections between brain hemispheres, so the brain presumably has greater ability to access and use information from both sides. You'd expect this to manifest as higher intelligence. This connection between hemispheres is particularly useful because, while they look like mirror images of each other, they are indeed functionally different, meaning they do different things.

The left hemisphere seemingly takes the lead with language processing,[74] while the right handles tone, pitch, and other fundamental sounds.[75] Studies also show an emphasis on global and local perception, in the left and right hemispheres, respectively, meaning the left hemisphere is more concerned with 'big picture' perception, while the right takes care of the fine details; the left brain sees the forest, the right brain sees the trees.[76]

So, the left and right side of the brain do indeed do different things, or similar things in different ways. And yes, people do usually have one hemisphere that is dominant, hence we're left- or right-handed.* There's also evidence suggesting that your dominant hemisphere influences your emotional capabilities.[77] Does this mean that the idea of the right hemisphere being responsible for all emotions is correct after all?

Not quite.

Back when brain-scanning technology was just becoming widespread, a growing body of evidence *did* support the idea that emotions are processed differently by the separate hemispheres.[78] Unfortunately, more advanced modern analysis and methods revealed that the situation is far more ambiguous.[79]

However, if you step back and look at it logically, given the size of the brain, how much goes on within it, how intensely interconnected it is, and how small and localised individual parts of the

---

* Each hemisphere controls the *opposite* half of your body: if you're right-handed, your left hemisphere is the dominant one, and vice versa.

brain can be while still having numerous diverse roles, attributing one specific function like emotions to an *entire brain hemisphere* is somewhat farcical. It's like insisting that everyone in the southern hemisphere on planet Earth is a great dancer, while everyone in the northern hemisphere can't dance because they're doing their tax returns. Such a claim would be ludicrous, and the same is true here, no matter how many memes and quizzes trumpet it uncritically.

So, if not one specific hemisphere, where in the brain *do* emotions arise from?

For a long time, emotions were believed to be the responsibility of the limbic system,[80] the brain region that essentially sits on top of the 'reptile brain'. The reptile brain, the label applied to the most primitive parts and processes of the brain (which have been around since dinosaur times, hence 'reptile', presumably) is actually the lowest layer of what's known as the 'triune brain'[81] model. This model proposes that the brain has three distinct layers, from the oldest at the bottom to the 'newest' and most sophisticated at the top.

The newer, smarter brain regions grew, evolved, out of the lower, more primitive ones, like a muffin, with a big bulbous top expanding out from the doughy base. Or like the rings of a tree, getting newer, and bigger, as you move from the centre of the trunk to the perimeter. But this tree is getting increasingly intelligent with each new ring.

As stated, the reptile brain is the bottom layer of the brain, responsible for basic physiological functions, like breathing, etc. The topmost layer – the huge wrinkly bit on top making up the bulk of the brain – is the cortex, or neocortex[82] (labels vary depending on who you're talking to). It's the 'human' bit of the brain, which does the impressive intellectual stuff.

Sandwiched between these two is the 'mammal brain', often referred to as the limbic system.[83] 'Limbic' is derived from the Latin word *limbus*, which means 'border' or 'edge', because the limbic system forms the border of the cortex, before the brainstem begins.

For a long time, the limbic system was believed to handle all the brain functions more complex than basic physiological processes, but more fundamental than the really sophisticated, intellectual stuff. Things like learning and memory, motivations and drives, reward and pleasure, conscious movement control, and, of course, *emotions*.[84] The higher, human brain, the top layer, the last bit to evolve, gives rise to 'conscious' things like analysis, language, attention, reasoning, and abstract thought.

The obvious conclusion here is that emotions are *sub*conscious processes. They're produced by the limbic system, a brain region which *pre-dates* consciousness as we know it, so they occur in the brain *below* consciousness, both figuratively and literally. Seems clear-cut, right?

Unfortunately, once again, it's not that easy, because the extent to which emotions occur consciously or subconsciously is another ongoing debate within the field of emotion research. A big part of this is the fact that the idea of a clearly defined limbic system that handles emotions (and more) is over 130 years old. However, in the light of modern evidence and our advanced understanding of the workings of the brain, it's another thing that's fallen out of favour. 'Limbic system' is still a widely used term for that general region of the brain, but the idea of it being a functionally well-defined, self-contained brain region is increasingly hard to support[85] in the face of the ever-increasing evidence revealing just how extensively connected everything in the brain is to pretty much everything else.[86]

One thing that particularly messes up the 'emotions must be a subconscious thing because they come from the limbic system' argument is that we now know that limbic areas have extensive two-way connections to the higher conscious regions, so both can influence and affect the other, in numerous ways.[87] Ergo, our conscious brain regions could easily be inducing our emotions, via their extensive links to the limbic areas. Many argue that this is exactly what happens.[88] The point is, even if emotions do *emerge* via the limbic system, we can't say for certain that they *originate* from there. It could be that this is like assuming all your letters are written by your postman. Again, it's an ongoing argument.

The widely accepted view today is that there is no one particular emotional 'bit' in the brain, no one specific section you can point at and say 'emotions come from there'. Instead, emotions are supported by a variety of networks or circuits,[89] where varied and widespread brain regions work together to create the experience of emotions that we all know and recognise (but struggle to describe).

This still doesn't really answer the question of where emotions 'come from' in the brain, and what processes give rise to them. For instance, a more modern view[90] is that emotions, and our reactions and behaviours induced by them, are processed by a circuit that includes the dorsolateral prefrontal cortex, ventromedial prefrontal cortex, orbitofrontal cortex, amygdala, hippocampus, anterior cingulate cortex, and the insular cortex.

It may sound quite specific, but these regions extend from the very top and front of the brain, where all the important cognitive work happens, right down to the very core of the limbic system in the centre of the brain, and encompass many areas in between. And this isn't even said to be an exhaustive list of the vital brain regions. Even if it were, all the named regions are known to have many diverse and important functions in other key processes too, like memory, attention, forward planning, pain perception, and more. They aren't *exclusively* involved in emotional processes.

On top of all that, even if a brain region is 100 per cent confirmed as having an important role in our experiencing of emotions, this often doesn't mean things are any clearer. A good example would be the amygdala, a small neurological region in the limbic system, in the brain's temporal lobe.[91]

For a very long time, the amygdala was best known for its role in processing and responding to the emotion of fear, and it's arguably still this function that the amygdala is best known for.[92] But as more data was accrued, the role of the amygdala expanded and diversified, and it's now known for its crucial role in providing the emotional component of memories;[93] for our ability to perceive emotions in others;[94] even for determining specifically which emotional response is required when we experience or perceive something.[95]

Far from having just one role in a single emotion (fear), the amygdala is now viewed as one of the key brain regions, a 'hub', even, for our experience of emotions.[96] The downside of this is that our grasp of how emotions work in the brain becomes more complex in turn.

So, while still a step up from saying a whole hemisphere is responsible for processing emotion, there's still ample room for ambiguity and uncertainty. As a result, 'Where do emotions come from in the brain?' remains an incredibly difficult question to answer, despite all our technical and scientific advances, and the reams of data generated over decades of study.

Part of that's undoubtedly down to there still being no real consensus on how to define emotions. If one lab is using a particular definition and another lab is using a different one, their results are less likely to match up, even if they're using the same methods. It would be like if two groups ran a survey of how many pets there are in the country, where one defined pets as, 'Cat, dog, rabbit, or goldfish', while the other defined pets as, 'Any non-human creature that lives in someone's home', so would have to include any vermin or spiders or termites too.

These two surveys would be looking for the same info, but because of their differing definitions (one being too specific, one too broad), they'd get wildly different results.

On top of this, even if we could specifically define emotions, the type of emotional experience being studied,[97] such as whether it's a pleasant or unpleasant emotion, will almost certainly have different expressions in the brain. I don't think anyone would dispute the observation that different emotions affect us in different ways.

It also depends if you're looking at the experience, or perception and expression of an emotion.[98] The human brain has a lot more overlap between these things than you might assume.

And that's not even considering the limits of the available technology for investigating this stuff. Based on the media coverage they get, you'd think brain scanners can read what's happening in your brain just like you or I would understand the images on a TV screen. Sadly, they're nowhere near that capable.

For example, current fMRI scanners, because of the indirect way they measure brain activity,[99] take several seconds to detect a change in such activity. But emotions happen *fast*. The processes underpinning them can be over and done in milliseconds, long before a brain scanner can figure out what's happening. Sometimes, using a brain scanner to study emotions is like trying to work out which horse won a race by going to the track three hours after it's ended and studying the hoofprints by the finish line.

Of course, this isn't to say that such studies have no value, because of course they do. It's just that there's still a long way to go. But for the sake of our (or, more pressingly, *my*) general understanding, perhaps 'Where do emotions come from in the brain?' is the wrong question to be asking?

A better approach might be to narrow it down, to look at individual and recognisable expressions and manifestations of emotions, and see what's happening in those instances specifically. Maybe this sort of approach will provide a metaphorical thread to pull on, which could help to untangle the greater ball of confusion that is emotions in general.

I hoped that was the case, because that's what I opted to do next.

2

Emotions Versus Thinking

I'm a big fan of science fiction.* But I'd be the first to admit that, if you consume enough, it can get a bit repetitive, with concepts and ideas that keep reoccurring. 'Despite having zero shared evolutionary history, alien races look a lot like humans with weird foreheads or ears' is one example of this. 'There is nothing so ludicrously dangerous that a shadowy corporation won't try to profit from it' is another.

A third is, 'Humans will always be threatened by, or otherwise inferior to, any intelligence that lacks or is immune to emotions'. The merciless artificial intelligences of the *Terminator* and *Matrix* franchises. The coldly efficient cyborgs like Robocop, or *Doctor Who*'s Cybermen. The intellectually superior Vulcans of *Star Trek*, for whom the rejection of emotions is the basis of their entire culture.† By accident or design, science fiction is regularly implying that our emotions are a liability, a weakness.

Admittedly, real life isn't much better. The Stoics and the Buddhists were insisting that emotions obstruct reason and enlightenment millennia ago. And referring to someone as 'overly emotional' is never a compliment.

So, the general consensus is that emotions are an obstacle to rational thought. It's like our brains have evolved beyond emotions,

* Shocking, right? Sorry for springing such a huge revelation on you without warning.

† According to *Star Trek* canon, Vulcans don't lack powerful emotions, but are able to suppress them almost entirely. This ability only slips during their seven-yearly 'pon farr' mating cycle. Or, whenever it's convenient for the episode's narrative.

but they're still hanging around, clogging up the workings of our minds: the psychological equivalent of an inflamed appendix.

I'd never put much stock in this idea before, dismissing it as the reserve of dystopian fiction, or posturing online pseudo-intellectuals. But when my dad fell ill, my inability to articulate or embrace my emotional responses was taking up far more of my headspace than I liked.

The severity of his condition fluctuated wildly from day to day too, so the emotions I was struggling to comprehend, or process, kept changing from morning to night. It was a challenge to get anything done. I really felt that my emotions weren't doing me any favours, just impeding my ability to think normally, to the point where the prospect of detaching my emotions, removing or shutting them off somehow, and allowing my thinking to progress unencumbered, became increasingly appealing. So much so, I ended up looking into how scientifically realistic it is.

And you know what? That's not how it works. At all.

It turns out that our emotions play many an intriguing, and vital, role in our thinking abilities, our perception, our minds. They may even be the reason we have those things to begin with. So, it was a good thing I didn't turn off my emotions. I could have done some serious harm.

Not that I ever really had that option. I'm a regular scientist after all, not a fictional one.

But if you want to know what to think about emotions, it's important to know the many ways in which the act of thinking pretty much *depends* on emotions. And that's what I'll explore in this chapter.

Emotivation

While trying to work out the emotions I was experiencing because of my father being hospitalised, I found I constantly wanted to *do* something. Anything! Like, for example, write about my emotions for a book. This one, that you're reading now.

This was a surprise to me. The traditional portrayals of sadness, anxiety, and grief, at least as far as I've noticed, suggest that they're very debilitating, leaving people bereft, or gripped with worry, unable to do anything useful. This can lead to the belief, or it did with me at least, that people experiencing negative emotions lack motivation. I'd argue that this is a reasonable assumption, given that 'lack of motivation' is one of the defining features of depression.[1] However, at a time when I should have been at my saddest, I instead experienced a strong urge to be as productive as possible.

Was this another sign that my brain was wired up wrongly in some way? Was I going to start singing show tunes whenever I tried to do maths, next? Or was it that I hadn't quite accepted the reality of my situation on an emotional level yet? Maybe my *rational* mind had grasped it, but my emotional processes were still throwing up error messages. Whatever the reasoning, I found myself with a lot of motivation, when I'd have expected to have none.

In truth, despite it being a big part of modern life, with companies and managers forever trying to motivate their workforce, and advertisers whose whole purpose is motivating people to buy certain products, few people appreciate just how complex motivation is.

Scientifically speaking, motivation is the cognitive 'energy' that makes us want to perform certain actions or behaviours. This may sound straightforward, but it manifests in countless interesting ways.

The urge to eat when hungry, drink when thirsty, to flee from dangers, to reproduce: these fundamental 'basic drives'[2] guide the actions of practically all species. And they're types of motivation. But the dedication required to spend years creating a great work of art, or building a successful business from nothing: that's motivation too. As is everything in between, from basic 'goal-directed' behaviours[3] where our actions are determined by the objective we want to achieve, to the desire to provide for our family and loved ones, i.e. people who aren't us.

Motivation is so complex because it's intrinsically tied both to our emotions and to our rational and logical conscious thinking processes (which, for ease of reading, I'm going to refer to as 'cognition' from now on). What we're ultimately motivated to do seems to depend largely on how emotion and cognition intertwine in our brains.

Looked at one way, it seems like motivation is more closely linked to emotion than to cognition. They're both derived from the same Latin word, *movere*, meaning 'to move'. And scientists have long accepted the link between emotion and motivation. Sigmund Freud himself described 'hedonic motivation', a classical approach which argues that we're motivated to pursue things that cause pleasure, and avoid things that cause pain.[4]

We are often guilty of doing things which are emotionally pleasurable but logically unwise. We've all been having a nice time and had 'one more drink' (or several) on a work night. This suggests that emotion is a more powerful motivator than cognition; because however much we may intellectually recognise something is beneficial, like heading home early and clear-headed, if it doesn't make us *feel* good, the motivation to do it is often harder to summon.

However, that's far from the whole story.

In the emotion science literature, the term 'affect' crops up repeatedly. When you're experiencing an emotion, you're in an 'affective state'. If you're researching how emotions work in the brain, you're doing 'affective neuroscience'. And so on.

Affect essentially refers to the experience of an emotion: what occurs in your body and mind when an emotion happens. All scientists agree that emotions do *something* to us. Affect is a way of referring to that 'something'.

Affect is made up of three distinct elements. One is 'valence', which is whether an emotion makes you feel good or bad. Valence can be positive or negative – e.g. happiness has positive valence, fear and disgust have negative valence.

Another element of affect is 'arousal': the degree to which an emotion stimulates us, mentally and/or physically. The mild frustration when a vending machine keeps your five pence change: that's low arousal. The intense fear and panic from nearly crashing your car is a very high arousal experience. Increased arousal usually corresponds to raised activity of the sympathetic nervous system.[5]

Finally, an affective state has 'motivational salience', or 'motivational intensity': the desire to act, to respond, that is induced by an emotional experience. Seeing something absolutely disgusting that compels you to look away has high motivational salience. The vending machine swallowing your change has low motivational salience.*

So, potentially *all* emotional experiences motivate us, to an extent. This is supported by evidence which suggests that emotion and motivation are processed by numerous overlapping systems in the brain.[6]

On the other hand, we don't constantly act on our emotions. We don't run screaming from *everything* that scares us, we don't persistently gorge ourselves on something we're craving, or lusting after. We may feel the urge to do these things, but we keep ourselves in check. We can do this because motivation, emotion, and cognition intertwine in interesting ways in the human brain.

Part of the brain which many consider to be the 'hub' of motivation is our old friend, the hypothalamus. Alongside its many other life-sustaining roles, the hypothalamus has a well-established role in motivation and behaviour.[7] The full situation is incredibly

---

* Unless, of course, it's just been one of *those* days.

complex, but in a sense the hypothalamus 'creates' motivation. It causes the urge to act and behave in certain ways, via numerous connections to the brainstem and other fundamental motor-control areas,[8] which are like the strings controlling the puppet that is our body. The hypothalamus is constantly pulling on them.

Research has revealed specific hypothalamic systems responsible for instinctive behaviours, largely concerned with eating, reproduction, and defence.[9] Absent-mindedly finishing off a bag of crisps while watching TV; finding yourself unthinkingly gazing at someone very physically attractive; immediately recoiling after touching something hot: these are all instinctive, reflexive behaviours. You're motivated to do them, but you don't *think* about doing them. You can thank your hypothalamus for that (or blame, in the case of the first two examples).

However, the hypothalamus is connected to *every* part of the brain,[10] so is not solely responsible for controlling motivation; all other parts of the brain also get involved.

Some parts are sub-cortical, limbic, emotional regions. Others are prefrontal and temporal lobe regions, i.e. cognitive ones. Both can modulate, or limit, the impulsive drives of the hypothalamus. Like, we can (and typically should) consciously stop ourselves from staring at an attractive person because we know, cognitively, that it's considered rude. Similarly, if we're feeling emotionally disgusted, the instinctive motivation to eat (aka appetite) diminishes.

This setup means emotional processes could lead to specific motivation, without any input from our cognitive brain regions. And vice versa.[11] We all occasionally do things out of sheer excitement, or fear, or rage, that we'd never normally do when we *think* about it. Conversely, we often complete household chores with scant emotional engagement; we're motivated to do them because we consciously know they need doing. Rarely do we have any emotional *urge* to do them.

In this sense, the motivation-producing hypothalamus is like the engine of a car, with our emotions and cognition in the front seats, one at the steering wheel, the other holding the map, both constantly bickering about who gets to do what.

However, even if we accept that emotions and motivation are fundamentally linked, and the former regularly produces the latter,* what we're eventually motivated to do is regularly determined by our cognition. The smarter, more recently evolved regions in the brain's frontal lobe, particularly those of the prefrontal cortex, grant us the gift of executive control.

Executive control is a general term for several functions,[12] including impulse control, problem solving, working memory, self-regulation and assessment, and more. Executive control is the ability to overrule our more primitive, animalistic traits, like emotions, and instead use reason and logic to guide our thinking and behaviour.

This is the 'intellectual' part of our psyche, and it has a substantial role in motivation. When we decide we're going to do something, it's far from a simple binary yes or no process. Many variables are considered, like the effort or cost required,[13] potential reward offered,[14] risk involved,[15] and more. All these calculations have distinct neurological processes underpinning them, and all feed into what we're eventually motivated to do.

Say you really like cupcakes. This would mean you're instinctively motivated to eat cupcakes whenever you see one. But if you see one at the other end of a flimsy rope bridge over an active volcano, you wouldn't automatically be motivated to go get it. Your executive control steps in, assesses things, and overrules any pleasure-seeking emotional motivation to retrieve the cupcake.

---

* Interestingly, it's rarely the other way around. It's very hard to motivate yourself to experience an emotion. We can't just 'decide' to be happy, despite all the memes and 'inspirational' messages that insist otherwise.

Of course, this scenario, like most others, has multiple emotion-inducing factors, and they don't automatically match up. Here there are emotions saying, 'Cupcake nice! Get cupcake!', but also emotions saying, 'VOLCANO HOT FIRE DEATH! AVOID!' But still, it seems it's the logical, cognitive systems in your brain that take these competing signals into account, and have the final say about what we do.

Evidence strongly suggests that the part of our brain responsible for integrating our emotional impulses into our rational decision making (and subsequent motivation) is the orbitofrontal cortex. While the orbitofrontal cortex's many functions are still being explored, it apparently plays a key role in self-control, particularly regarding emotional motivations.[16]

For instance, if you see a sexually attractive person at a party, you might experience lust: an instinctive, emotional drive to have sex with this person. So, you are emotionally and instinctively motivated to engage in behaviours likely to achieve this goal.*

However, you're surrounded by people you know, one of whom is married to the person you're lusting after. 'Making a move' *may* lead to a positive emotional experience, but the negative emotional consequences (e.g. social ostracisation, the destruction of valued relationships) considerably outweigh it. So, you're actually motivated to suppress or ignore your sexual urges, not obey them.

And it's your orbitofrontal cortex which allows this to happen. It weighs up the pros and cons of emotional desires and determines whether they're worth doing. It's the proverbial angel on our shoulder, constantly saying, 'Are you sure about this?'

At the most basic neurological levels, motivation is expressed in terms of 'approach' or 'avoid'. In the earlier 'cupcake over a

---

* Although what these are will be very subjective, and vary wildly from person to person.

volcano' scenario, you can either approach, or avoid, the cupcake. Obviously, you avoid. But this applies to countless, far more mundane scenarios, too.

Do you approach or avoid the kitchen sink piled high with dirty dishes? Sometimes we sigh and pull on the rubber gloves. Sometimes we scarper and hope someone else in the house needs a clean plate first. Approach motivation and avoid motivation has won out, respectively.

Research shows that our emotional state plays a big part in whether the approach or avoid system dominates. So, in this case, cognition is influenced by emotion, not the other way around.

Take the way that anger can fire us up, motivating us to have it out with a jobsworth bureaucrat, or to yell expletives at the neighbours for playing loud music at 2 a.m. (Again!)

Or, if tackling the exact cause isn't possible, anger compels us to do *anything at all* to relieve the pressure. People punch walls, scream into pillows, bellow abuse at anyone unfortunate enough to wander into the room, however innocent they may be.

It's frequently unfair, and rarely logical, as the Stoics recognised 4000 years ago, but anger is certainly very motivational. It makes us want to do things, regardless of risk, effort, or reason.[17] And that's because it significantly raises activity in the 'approach' motivation system, in the prefrontal cortex.[18]

Fear does the opposite. When experiencing fear, we're far more likely to avoid things.[19] The sound of a twig snapping is of no consequence when happily walking through a sunlit park. But hearing the same sound while creeping through a dark forest in the dead of night makes us want to desperately flee. You're strongly motivated to get away from whatever made that sound, even if there's no particular rational reason to do so. Because you're experiencing fear.

The important role of emotions is also revealed by the different potencies of what's labelled extrinsic and intrinsic motivation. Extrinsic motivation is where we do something because someone or something is 'making' us do it, either by offering rewards (e.g. your employer pays you if you come to work), or promising punishments (e.g. your employer will fire you if you don't come to work). Intrinsic motivation is where we're motivated to do something because we've decided it's something we *want* to do, that we'll enjoy or benefit from.[20]

An artist painting a picture because someone is paying them for a commissioned piece is extrinsically motivated. The same artist painting a picture because it's something they want the world to see is intrinsically motivated. Both types of motivation can often apply to the same thing.*

However, evidence suggests intrinsic motivation is the more potent and enduring. A 1973 study[21] rewarded one group of children for playing with arts and craft materials, while another group were given the materials and left to do what they wanted. When investigated at a later point, children who were rewarded were *less* motivated to play with the same materials, compared to children who had enjoyed them on their own merits the first time around. The dominance of intrinsic motivation has been recognised ever since.[22]

Indeed, quitting an uninspiring job that paid the bills and embracing a far more financially precarious existence, to 'live the dream', is the backstory of countless performing artists, and a prime example of how much more motivational intrinsic factors are, compared to extrinsic ones.

Intrinsic motivation clearly arises when things stimulate us on an emotional level. If something is our passion (an old label

* I say this as someone who makes a living by writing about his interests and passions.

for emotion), we're often motivated to pursue it for years on end, with no obvious or guaranteed payoff. There's no rational, objective reason for doing it, beyond it being emotionally rewarding.

Many corporations have seemingly cottoned on to this. How often do we now wander into somewhere like Starbucks to find ourselves surrounded by posters and branding telling us that we're 'part of the family'? They're not just offering a caffeine hit, but an emotional connection!*

Undeniably, our emotions, cognition, and motivation interact all the time in our brains, in very convoluted ways. Untangling them is the focus of much research, particularly in the area of education and learning.[23]

One such person doing this kind of research is Dr Chris Blackmore of Sheffield University, who studies the role of emotional elements in online learning platforms.[24] I asked him what the latest understanding was regarding how emotions and motivation interact:

> There seems to be increasing awareness that the idea of positive and negative emotions as being good or bad, respectively, for learning, is too simplistic. I certainly found with e-learners that so-called negative emotions, such as frustration or anxiety, often preceded a breakthrough and transformation.

This was intriguing. For all the tropes of people pursuing their dreams to do what they love, what makes them happy, apparently the things that frustrate and stress us can be just as motivating. Wanting to *avoid* something that does, or could, cause the experience of distress or discomfort (emotional or otherwise, one assumes), can be a powerful motivator to 'do' something.

---

* Personally, I find it a bit much. You just want a coffee after all, not to be adopted.

This offered an explanation for my bizarre desire to keep busy during my father's illness. It's not that I was denying what was going on in my life, but the intensity of the negative emotions at play affected my motivational processes, raising activity in them so I was compelled to do something, *anything*, to avoid the discomfort of what was going on.

Far from being debilitating or disruptive, though, Dr Blackmore informed me that this phenomenon has had some very profound outcomes. Many of history's great philosophers and thinkers weren't necessarily motivated by a passion for discovery, or love of knowledge, but by a sort of existential dread.[25] Not knowing something so fundamental and important as how the world and our lives worked? That *worried* them.

Uncertainty is something the human brain really doesn't handle well. People often say, 'The waiting is the worst part', and studies have indeed shown that not knowing whether an unpleasant outcome will occur can induce more stress than the outcome itself,[26] which, however negative, at least offers certainty and clarity.

Basically, the great philosophers, responsible for some of the most profound realisations in history, were motivated by a form of *fear*. Dr Blackmore summarised it nicely:

> I reckon Kierkegaard had it right when he said 'whoever has learned to be anxious in the right way has learned the ultimate'.[27] Given how we so regularly assume that emotions obstruct logic and reason, it's weird that some of our greatest thinkers were so motivated by them.

But then, these iconic philosophers lived a long time ago, when religion and superstition held far more sway. Maybe that's why their motivations were less than 100 per cent rational? Would an equivalent modern-day thinker be similarly beholden to emotional factors?

To answer that, I sought out such an equivalent. These days, figuring out the universe and all it entails is the work of particle physicists, astrophysicists, and cosmologists. People like @AstroKatie, the Twitter alter ego of Dr Katherine 'Katie' Mack, astrophysicist and assistant professor at North Carolina State University. Dr Mack is a prominent science communicator, and author of *The End of Everything*, about the ultimate fate of the universe itself.* She told me:

> I often get messages from people who want me to reassure them the universe isn't going to end at any moment. The thing is, as a physicist, I can say it's very unlikely . . . but can I totally guarantee it? No, I can't.

How the universe will end is arguably the biggest question in modern science, so I wanted to know what her motivation was for taking it on. In response, Dr Mack told me about her moment of epiphany:

> I was an undergraduate at the regular dessert night for astronomy students. So, we're at this professor's house, and he's giving us tea and cookies and talking about cosmic inflation.† Specifically, about how the early universe expanded at an accelerating rate, which totally shaped the cosmos as we know it. He pointed out that we don't know why this accelerated expansion started, and we don't know why it ended. So, there's nothing to say it couldn't happen again. Right now.

---

* I'd say she's a 'star' of the science world, but that's probably not a compliment to an astrophysicist. It might be like calling a builder a brick.

† Standard social gathering stuff, when you're a scientist.

My own scientific knowledge pretty much begins and ends within the confines of the human skull, so the revelation that the whole universe could suddenly start behaving very differently was quite a profound one. Noticing this, Dr Mack made an appreciated effort to explain it in more small-scale terms. That these still involved planetary annihilation says a lot about how astrophysicists think.

It's like when you see evidence, like ancient craters, for a meteor strike. For me, it makes it clear that there are big things out there that have happened, can happen, that *do* happen, that can seriously alter my life, my environment. And I have no control over that, I'm just this tiny speck, clinging to a rock, and all these factors that I think of as very solid are at the whim of cosmic forces. That kind of stuck with me.

Assuming Dr Mack is a reliable representative of her field (and evidence suggests she is), it seems those investigating the fundamental questions of existence itself are still motivated, at least partly, by a sense of anxiety, concerning how our universe works.

We are, at present, laughably helpless to do anything about the fate or behaviour of the universe. That's not a comfortable feeling, if your rational mind is inclined to contemplating such things. Trying to reduce the uncertainties of our existence and how it functions won't change that, but it *can* instil a sense of control, of autonomy, however minor or inconsequential, which helps reduce the anxiety.[28]

But then, perhaps I'm reading too much into it? Who knows why such titans of intellect do what they do? Emotions may play a part, but presumably those who plumb the depths of the cosmos for answers are far more reliant on cognition than emotion.

But then Dr Mack said this:

In the course of researching my own book, I spoke to a bunch of different cosmologists. And I always asked 'How does the end of the universe make you feel?' The idea that there will be a 'heat death' of the universe, that everything fades to black, many found that really depressing. Some even said 'I just don't believe it's going to be like that', and have since produced their own alternative theories and ideas now, because they simply *do not like* the idea that the universe is going to fade away and die.

So, numerous extremely intelligent people, if faced with mountains of data and peer-reviewed evidence which says how the universe will end, will reject it. Because it's too depressing. Too bleak.

Dr Mack did clarify that her esteemed colleagues weren't *just* motivated by an emotional dislike of how the universe may end, that their arguments and alternative theories were based on actual data. There is, admittedly, a lot of uncertainty when you're studying something trillions of years in the future. But that emotion helps shape their investigations, and motivates them to look for alternatives, is hard to get away from completely.

It turns out that, even in the most cerebral settings, our emotions can still motivate us. In certain contexts, emotions can *alter the fate of the universe*. Or, at least, our models and theories about it. Perhaps we should start being more respectful of emotions?

Having said all that, as grand a scale as it is, seeing how the cosmos pans out is still a largely theoretical affair. So perhaps it's not surprising that emotions, with their ability to affect our thinking, can influence it like they do?

By contrast, surely our emotions can't affect how we see the real, tangible environment right in front of us? You'd think so, but you'd be wrong. Our emotions do indeed colour our perception of the world around us. And I mean that very literally.

The colour of emotions

When most of your day is spent worrying about a parent, it inevitably means you spend more time thinking about your childhood and upbringing. That's when your parents' presence in your life was at its most prominent, and important. But when your brain is constantly dredging up random memories from your youth, it'll eventually cough up some of the more bizarre and surreal stuff you experienced.

In my case, it happened as I was about to wash up after dinner. A particularly bizarre memory hit me, as I was staring at a pack of brightly coloured dish sponges kept under the kitchen sink.

When I was almost eighteen, the oldest member of my friendship group moved out of his parents' house and into his own. He promptly invited everyone else in our extensive group to come over, and so we both did.

For context, at the time we were teenagers in a tiny, isolated South Wales former mining community in the late 1990s, before smartphones and the internet; our social lives mainly involved hanging around in each other's homes, which meant we had to put up with ever-present parents, constantly reminding us we should be studying, or overhearing our increasingly explicit conversations (we were male adolescents, and testosterone is potent stuff).

So, one of us having his own place, where we could say and do what we wanted, without being nagged or yelled at? That was ideal!

However, as soon as he'd moved into his new place, my mate had, for some reason, repainted every room a vivid primary colour. The front room was glaring purple. The lounge, stark orange. The kitchen, almost-neon green. The bedroom, fire-engine red. It was like the lair of one of the campier *Batman* villains. If there was a clown-themed torture chamber in the basement, nobody would have been surprised.

I don't mean to criticise my school friend's interior design choices, but it's tricky to relax with a few drinks when you're getting hangover symptoms from the décor.

Why does that happen, though? At the end of the day, colour is just photons of certain wavelengths hitting my retinas.[29] How does something so fundamental trigger a potent *emotional* reaction?

In truth, colours have interesting, and surprising, effects on our brains. They influence our emotions, and our thinking in turn. There's a whole discipline, colour psychology,[30] dedicated to studying how and why certain colours affect us.

As mentioned earlier, we humans (and other primates) are trichromatic: our eyes can detect the three colours of red, blue, and green. But some species, which have developed under different evolutionary pressures, can't see colour at all. Others – typically birds or marine creatures – can detect four, five, or more colours. The current record holder is the mantis shrimp,[31] with eyes that are sensitive to a frankly ludicrous *twelve* different colours.

My point is, while colour itself, the result of the wavelength of photons, may be (relatively) simple, our ability to perceive and recognise it is anything but. It's down to complex systems in our brains, ones that evolved and developed over millions of years.[32] This means there's ample scope for the neurological mechanisms of colour perception to be intertwined with the brain's emotional systems.

For comparison, consider the road network and the sewer network of a modern city. Even though they have very different purposes and work in completely different ways, they exist alongside each other. And although they usually operate independently, they undeniably can, and often do, influence each other. The growth or alteration of one must take the other into account. And if there's an eruption in a sewer running under a road, the road users are certainly affected by it.

But the evidence suggests that the colour vision and emotion processing mechanisms in our brain are even less 'distinct' than that.

Vision is the dominant human sense. Some estimates suggest 80 to 85 per cent of our perception, learning, thinking, and general brain activities are mediated through vision in some way.[33, 34] So, the idea that seeing certain colours would trigger an emotional response isn't such a stretch.

That may be why we regularly describe emotional experiences in terms of colour. Sadness leads to us 'feeling blue'. Fury is associated with 'seeing red' or being 'red with anger'. Coveting another's possession or attribute leaves us 'green with envy'. And so on.

These associations could be learned or cultural things, admittedly. Maybe some historical artist depicted an angry person as red for purely aesthetic reasons, this caught on, and the association has persisted ever since.

However, while such cultural factors do undoubtedly play a part, evidence suggests these colour–emotion associations are more fundamental, more 'natural'. For one, they appear to be surprisingly consistent across cultures.[35] Considering the huge differences in history and development, there's more cultural agreement regarding which emotions are associated with which colours than you would probably expect.

Red is the most widely studied colour in this context.[36] Evidence shows that people regularly associate red with anger[37] and/or danger (i.e. fear).[38] Other colour–emotion associations, like blues and greens being 'cool' or 'calming', have been demonstrated repeatedly too.[39]

There are many theories about how these associations might have evolved. If our primitive ancestors saw spilled blood it likely meant a predator had been, or *still was*, nearby. Hence, we learned red means danger. Maybe green means disgust because harmful rotting things often turn green due to mould and decay. Some even argue that blue is associated with sadness because we cry when we're sad, and tears are water, and water is 'blue'. A bit tenuous, but technically I can't rule it out.

However, one particularly interesting possibility brings matters back to our old friend, the face. Some studies suggest that primate colour vision is particularly sensitive to the range of colours produced by changes in blood flow to the skin of the face.[40]

If we're too hot, blood is shunted towards the skin, to expel internal body heat. So, we look red. Conversely, when we're cold, blood is shunted away from the skin, to minimise heat loss. Due to the physical scattering of light, the chemical composition of deoxygenated blood, vessel constriction, and visual processing,[41] this makes our skin appear blue. Or blu*er*, at least.

It's not just temperature; emotions have this effect too. Some emotions mean high-arousal, high-energy states, meaning we 'go red', be it via angry flushing or embarrassed blushing.[42] Others, like fear, direct blood to our important internals, to get ready for fight or flight, so we go pale white/blue as blood leaves our face. Essentially, our emotions change the colour of our face.

It may seem like this facial colour change is just a coincidental by-product, like how you end up with a yellow patch of grass on your lawn if you leave a paddling pool on it for several weeks. You don't have any use for a yellow circle of lawn. It wasn't planned. But you've got one anyway.

However, this facial colour changing may be far more important than just an accidental result of other processes. Firstly, while humans clearly have less body hair than other primates, practically *all* primates go for the 'hairy body but bald face' look.[43] All other hairy creatures have hair covering their faces too, but not our evolutionary cousins and us.

Clearly, seeing the bare skin of the face is important for primates. And while we do use our facial expressions to convey a great deal of information, you technically don't need *bare* skin to do that. The only information hairless skin adds to an expression is . . . changes in skin colour.

There's also, as previously mentioned, data showing that primate colour vision is particularly sensitive to the different colour shades produced by variations in blood flow to the skin.[44] This implies the colour of our face conveys something very important. But what?

Well, some studies suggest that certain areas of our face, around our mouth, nose, and eyes, change colour in specific ways according to the emotion we're experiencing. A 2018 University of Ohio study[45] reported that subjects could tell what emotion a 'neutral' face was displaying from the colour patterns applied to it. The implication here is that specific emotions are represented by specific patterns of colour on our faces.

Some argue that seeing these emotional facial colour patterns is why primates evolved bald faces and elaborate colour vision. The profound implication here is that, rather than being unlikely or coincidental, a fundamental association between certain colours and certain emotions is the whole reason we can see colours *at all*!

An interesting notion, but as ever, there are some issues. For instance, not every human has the same skin colour. How does that factor here? Although, this has been studied, and the effect seems to persist, somehow.[46]

Or maybe this conclusion is backwards: maybe we evolved faces that displayed emotions of certain colours because our species was best able to see those colours? Also, this theory adheres to the classic 'our facial expressions directly correspond to our primary emotions' stance, and we know that's not as unshakable as many believe.

But nonetheless, our brains do seem to instinctively associate certain emotions to certain colours. And this can have some weird effects.

For instance, if exposed to a specific colour, it can alter our perception of how hot, or how loud something is.[47] Also, people recover faster from stress, mental fatigue, even physical injuries, if they're exposed to natural, leafy, *green* environments.[48] Studies into this phenomenon (known as 'attention restoration') suggest that it can work even if you just use the colour green, divorced from the context of nature.[49]

Blue is regularly deemed a calming colour (depending on shade). Maybe that's why doctors and hospital staff normally wear green or blue scrubs, or neutral white: to help calm and reassure sick patients, who are likely to be (understandably) worried.

Conversely, you never see a medical professional in bright red clothes (unless a surgery has gone spectacularly wrong). Red is strongly linked to feelings of anger, danger, and threat. Many warning signs are red, regardless of what they're warning against.

These instinctive colour–emotion associations can manifest in very weird ways. Several studies have shown that wearing red can boost your chances of winning at competitive sports.[50] How? Possibly because we fundamentally link red to threats, so our brains instinctively divert more attention to something if it's red. And when you're playing a demanding, fast-paced competitive sport, any distraction at all can make a difference.

Researchers often refer to this process, where perceived threats divert attention from a task, as 'goal distraction'.[51] Amusingly, one study reported that footballers score fewer penalties against a goal-keeper wearing red,[52] so goal distraction led to *literal* goal distraction.

Maybe that's why things like clashing colours and my friend's eye-watering interior design inspire such negative reactions; it's not the colours per se, but the presentation. They're so bright, and/

or violate patterns and expectations to the extent they take up too much attention, meaning we struggle to focus or relax. Our brains don't like that.

Before we get carried away, it's important to acknowledge that colours alone don't dictate our emotional reactions. Our brains are way more sophisticated than that, and our development and experience and environment and context will all have a big part to play, which makes the overall picture a lot more complex.

For instance, yes, red is regularly associated with anger and threat. But also, sexual arousal. Also, warmth and cosiness. Quite an impressive range for a single shade. And nobody argues that Santa Claus wears a red suit because he's constantly furious. My point is, there are plenty of other factors that determine what emotions we experience, not just colour.[53]

But, based on everything covered here, the fact that colours *can* have fundamental emotional and cognitive effects in our brains is increasingly hard to deny. So, I won't deny it. Not as much as I once did, anyway.

One thing about this did strike me as odd, though, about all the data showing that red is associated with danger, threats, and aggression. Even if all that is true, people *like* the colour red. It's very popular. So, countless people experience a *positive* emotional association with something that should be inducing a *negative* one.

That's not how emotions are meant to work, is it?

. . . is it?

So bad, it's good

When my father fell ill, I started watching sad movies, to sort of trick myself into crying, in an effort to clear the 'emotional logjam' I was experiencing. Specifically, Pixar movies. My wife and I are long-time fans in any case, and they seem especially adept at delivering an emotional gut punch.

This approach worked for a bit, then hit a snag. Remember, at the time we were under lockdown because of the pandemic, so I was stuck at home with my kids. My daughter likes watching things with me, but she was only four, so interested more in bright colours and fun scenes than character arcs and plot developments. She'd be clapping and cheering at the colourful balloons in *Up*, or the rainbow wagon sequence in *Inside Out*, only to turn around and see me weeping. At something she thought was great fun.

I feared I was fundamentally confusing my little girl during an already stressful period. So, I decided to look for alternative options for jump-starting my negative-but-necessary emotions. And I was spoiled for choice. There's a glut of movies, TV shows, books, articles, and music out there, designed to make you feel sad. Or angry. Or scared. Even disgusted.

If anything, entertainments and artworks that provoke these typically avoided emotions are more respected than those that make you feel, you know, *good*. It's not that nobody's ever won an Oscar for making people laugh, but it's a relatively rare occurrence. But make enough people cry with your performance, they'll be queuing up to hand you statuettes.

And I couldn't help but wonder . . . why? Why are things that cause ostensibly *negative* emotional reactions so counterintuitively popular?

I previously mentioned valence,[54] the affective (emotional) property that states whether an emotion is positive or negative. Most would agree that certain emotions usually make us feel either better or worse. Except that can't be the whole story, or we surely wouldn't actively seek out experiences that stimulate negative emotions. So, why *does* the human brain end up liking things and experiences it technically shouldn't?

Much of it is down to the ways our emotions and cognition interact. The most obvious example of people liking something objectively negative is the worldwide popularity of spicy food.[55] Capsaicin, the chemical found in chilli peppers, triggers receptors in the nerves of the tongue. Some of these detect temperature, hence we think of spicy food as being *hot* regardless of its actual temperature (a jalapeño eaten straight out of the fridge will still taste as fiery).

But spicy food doesn't just taste hot; it *burns*. Anybody who's ever been chopping raw chillies then rubbed their eye, scratched their nose, or, god forbid, used the toilet, will be very aware of this. It's because capsaicin also triggers nociceptors, the receptors in nerves that transmit pain.[56]

Why would we humans so persistently enjoy eating actual pain? Much research has been dedicated to this question. Many possible answers have emerged, like historical practices where chilli was added to food because of its antibacterial properties,[57] or thrill-seeking tendencies in humans,[58] or male dominance behaviour and self-assertion.[59] Overall, there are many potential factors leading our brains to enjoy the experience of literal pain, ranging from the most fundamental, biochemical levels (e.g. quirks of DNA and brain development) to the more cerebral and abstract (e.g. culinary traditions of our culture influencing our preferences).

It does seem quite clear that we aren't born liking spicy food, though. It's an acquired taste; we grow to like it over time, hence spicy baby food has never been a thing.

Speaking of acquired tastes, another area where people seem to actively enjoy 'unpleasant' sensations is BDSM, i.e. Bondage, Discipline, Sadism, and Masochism, the sexual practice where people enjoy inflicting pain, restraint, or humiliation on a willing partner, or having it inflicted upon themselves.

Despite being a completely consensual arrangement between (typically enthusiastic) partners, BDSM is regularly treated with disdain or suspicion in the mainstream. Yet, appropriately enough given the 'liking things we shouldn't' aspect, there's always been a notable public fascination with it too, as the success of *Fifty Shades of Grey* demonstrated.* This is undoubtedly linked to how BDSM provides another stark example of people enjoying things that literally cause pain.

Because of this, science has long been interested in BDSM too. And it's caused a rethink of our existing notions of how pain works in the brain.

Our brains have evolved a sophisticated pain management system, which involves releasing endorphin neurotransmitters in relevant areas, to cancel out pain, providing pleasure and relief instead.[60] Endocannabinoid neurotransmitters† perform a similar function.[61] The upshot of this is that, if done right, pain can lead to pleasure.

That's particularly true for human sexual acts. While incredibly diverse, even the most 'vanilla' expression of human sexual activity is intensely, intimately physical. Consensual sex, of any sort, can easily cause pain, however unintentional.

Luckily, during sexual activity, our brains modulate our perception and processing of pain, via areas like the periaqueductal grey.[62] Sex is fundamental to the survival of our species, but if it were constantly painful nobody would do it. So, the pain felt during sex is very different to pain experienced at other times.

Our brains essentially make pain *enjoyable* during sex. The initial sensation is processed differently, so that it enhances, rather than

---

* However, the BDSM community insist that *Fifty Shades of Grey* doesn't feature true BDSM at all, but rather a toxic relationship between a woman and a sociopathic billionaire who enjoys hurting people. I've never read it, but if people who enjoy being whipped find a book intolerable, that's a bad sign.

† Endorphins are the brain's own opiates (morphine, heroin, etc.) while endocannabinoids are cannabis equivalents. Those drugs only work because they stimulate or hijack these pre-existing systems in the brain.

disrupts the experience. It's like how raw meat, to us modern humans, is actively unpleasant, and even dangerous to consume. But if you cook it, it becomes the opposite. Same substance, same components, but *processed* differently.

Does this explain the appeal of BDSM? In part, maybe. But there's more. Human sexual behaviour encompasses far more than just the purely physical act of intercourse; there's typically a strong emotional element too. When it's lacking, sex can be very unsatisfying, even upsetting.

BDSM has a very potent emotional component. Participants are typically submissive, or dominant; they like to be hurt, or do the hurting, respectively. To make sense of this, consider that interacting and forming bonds with other humans genuinely causes pleasure via the reward pathways of our brains.[63]

Another thing we instinctively respond to is status. Raising our social status, being better than others, elicits a positive emotional reaction (happiness, satisfaction, pride, etc.).[64] Similarly, low social status causes serious stress and anxiety, even in non-humans.[65]

BDSM seems to heighten all this. Studies of BDSM enthusiasts[66] reveal that submissive types experience heightened pleasure throughout the experience. They're surrendering absolute control over their very bodies to someone else; it's hard to imagine a more intense interpersonal bond than that.

By contrast, those who are dominant seemingly enjoy BDSM only when there's an element of 'power play', where they have complete control over their submissive partner. Such superior status over a submissive partner is presumably very pleasurable. But also, the complete trust given – to be allowed such direct control over someone's physical and mental wellbeing – must be quite a rush too, for such social creatures as us.

So yes, BDSM has a *very* strong emotional element. Physical sexual contact is a surprisingly small part of BDSM, with enthusiasts regularly confirming that the main source of pleasure is the emotional bonding and experience.[67]

Ultimately, the brain's reprocessing of pain during sex isn't sufficient explanation for the appeal of BDSM, because sex often *isn't happening*. Maybe the intense emotional experience overrides the pain? Or maybe they combine in some interesting way, producing a wholly new experience?

Indeed, some studies suggest the experience of BDSM can lead to 'altered consciousness', akin to that experienced during mindfulness meditation.[68] Odd to think of BDSM enthusiasts as modern-day monks, but consider how many religions have embraced aspects of torture or self-flagellation.[69] The link between pain and enhanced consciousness may be a very old one. The BDSM community just have more fun with it.

But however intriguing spicy food and BDSM may be, visceral pain is *not* an element of most negative or unpleasant emotional experiences. And yet, people still regularly take pleasure in works that induce them. Clearly, there's more occurring here.

For instance, one explanation for enjoyment of extreme sports, horror movies, or anything that scares us (despite the whole point of fear being to keep us *away* from whatever causes it), is the excitation transfer theory.[70]

It can't be denied that fear is very stimulating; the fight-or-flight response ramps up the whole brain and body, putting us in a heightened state of awareness, the better to cope with the imminent danger. And this state doesn't vanish once your parachute lands, or you turn the slasher movie off; it endures for a while. Anything you experience while in that heightened state is, therefore, more stimulating, more exciting, more . . . enjoyable?

There's also relief that the cause of the fear is gone. The removal of a bad thing can be just as rewarding as the presenting of a good thing[71] when it comes to how our brains learn what behaviours and actions are to be encouraged, repeated. Which would explain why fear-inducing activities and entertainments keep people coming back for more.

Another thing putting a positive spin on negative emotions is *novelty*. Like many species, humans inherently like new things (as long as they're safe). Our brains automatically learn to ignore anything too familiar, too predictable.[72] Therefore, novel experiences are more stimulating. Novelty increases activity in the pleasure-producing parts of the brain.[73]

Consequently, humans are invariably drawn to new experiences. Everybody's bucket list is made up of things they've never done before. Nobody wants to complete one last daily commute before they die.

Because we regularly *avoid* things that cause them, a negative emotional experience is typically more unusual, more novel, than a positive one. So, we can gain some minor reward from an unpleasant emotion purely because it's atypical.

Moreover, negative emotions can be genuinely beneficial once our cognition gets involved. Ever had a random thought about doing something bad? Don't worry, you're not a psychopath; our consciousness is regularly throwing up unpleasant or alarming scenarios, like 'What if I jumped off this cliff?' 'What if I stole that money sticking out of a stranger's pocket?' 'What if I set fire to that abandoned house?' And so on.

We *know* they're wrong, these intrusive thoughts.* They make us feel bad. But we seem powerless to stop them. That's because they're *useful*. The negative emotional reaction they provoke reinforces our ideas of what's wrong or right; it tells us we're correct.[74]

---

* Sometimes referred to as 'forbidden' or 'taboo' thinking/thoughts when they're about doing things we believe are wrong. The term 'intrusive thoughts' usually applies to all the idle speculations or inane thinking that our brains churn up regularly, not just the darker stuff.

It's like our brains are a well-guarded fort, where soldiers are sent out to patrol the defences and check for weaknesses, maybe even stage a mock attack to keep everyone on their toes. Thinking intrusive thoughts and having the expected emotional response is the brain's way of checking that its understanding of how things work is still robust and reliable. And that's helpful.

But once again, the cognition–emotion relationship is a two-way street, and our emotions have many pronounced effects on cognition. For example, one persistent finding is that positive emotions broaden our cognitive scope, while negative ones narrow it.[75, 76] To translate: when we're in a positive emotional state, our brains tend to take everything in, not concentrate on one particular thing. By contrast, when we're in a negative mood, we concentrate more on one specific thing at a time, pay more close attention to whatever we're dealing with. It's 'big picture versus fine details' again.[77]

In a sense, if our cognition was a theatrical production, our emotions are the lighting. Positive emotions bring the house lights up, so you can see every actor, prop, and backdrop. Negative ones operate spotlights, so our attention is only focussed on those performers and set illuminated by the beam.

This sounds like positive emotions are better, but it's not that simple. Our brain's attention capacity is rather limited;[78] spreading it too wide means we miss things, and end up relying on what's already in our brains, like prior experiences, and established beliefs and understandings.

Unfortunately, our prior experiences and understandings may well be incorrect, or irrelevant to our current situation. People in a good mood, a positive emotional state, are seemingly more prone to making errors, like blaming the wrong person for something, being too gullible, even resorting to racial stereotypes or other prejudices.[79]

To put it another way: being happy might *feel* nice, but seems to hamper your ability to *be* nice. Or at least, to be focussed. This explains the data which shows that happy employees aren't nearly as useful as current corporate thinking suggests.[80]

Negative emotions, however, make you *more* focussed, meaning you take more time and dedicate more neurological resources to the decisions required in certain situations.[81] This explains why negative emotions make us less gullible, less discriminatory, cause us to form better judgements about others, have better recall of events, communicate better, and so on.[82, 83, 84] This all makes sense if negative emotions simply make you pay closer attention to what's going on, basing decisions and actions on the details of the situation you're in, rather than assumptions and prior experiences.

One explanation for why this happens is negative emotions producing activity in our threat detection systems, meaning they heighten focus via the same neurological mechanisms as hazards and dangers.[85] Maybe all these (surprisingly helpful) cognitive impacts of negative emotions explain the common link between suffering and creativity,[86] why so many great artists and thinkers are such troubled souls.

But it's not just the indirect effects they have on other neurological processes; negative emotions are important in their own right, *vital* even, for good mental health and wellbeing.[87]

An emotional experience doesn't immediately disappear from the brain. Just like how the food you eat doesn't vanish from your body once you swallow it, or the pain of being hit doesn't stop as soon as the impact is over, the memory of an emotion, and all the effects it had, can linger. And for a very long time if it was particularly potent.

We've already seen how many different areas and networks and processes in our brain are involved in emotions. An emotional experience is felt throughout the brain, so, logically, it has the potential to cause substantial changes in the brain's usual operations. Consequently, the effects an emotion has on us, on our brains, don't automatically fade like breath on a mirror. They may need to be worked out, and processed.

'Processed' is the key word. Someone who's experienced tragedy or trauma is said to 'need time to process things'. The grieving process is perhaps the most familiar example.[88] Emotional processing[89] is where an emotional experience, and all its neuropsychological elements, are integrated into your brain's existing setup, so it can resume normal functioning (or as close as possible).

If your brain were a busy office workplace, a profound emotional experience is like a new employee sent to work there. It's nothing unusual, but said new employee needs a desk, a company ID and network account, duties and assignments, etc. It's standard procedure, but nonetheless takes time and effort.

Similarly, our brain may need time and resources to effectively incorporate emotional experiences. Usually it's so quick and efficient we don't even notice. Our everyday emotional experiences just slot into our psyche; they're not new employees to the office, more like existing employees wandering in and out as they do their jobs. Nothing to worry about.

But a powerful, unfamiliar emotional experience? Particularly a negative one, like losing a loved one, or going through a violent incident like a major house fire, or natural disaster? That's like a new employee who turns up unexpectedly, doesn't want to be there, resents working at all, and only has the job because the managing director of the company is friends with their uncle. It takes substantially more time and effort to integrate them into the workplace. But it must be done, otherwise they'll just be standing about, getting in everyone's way, complaining, and generally disrupting the whole workplace, all while claiming a salary.

And so it is with the brain: strong, unprocessed emotions can cause problems with the overall operations. This explains why incomplete or unsuccessful processing of emotional experiences can lead to mental health problems, particularly post-traumatic stress disorder (PTSD).[90] The traumatic emotional experience that isn't properly processed, that maybe *can't* be processed in the normal way, causes disruption, throwing the (typically much smoother) integration between cognition, emotion, perception, behaviour, memory, etc. completely out of whack.

Tellingly, most psychological therapies for PTSD involve some means of approaching the cause of the trauma (or the memory of it) in alternative ways, that *don't* trigger the debilitating fear and anxiety.[91]

To switch metaphors for a moment, think of our normal brain functions as a vital road that goes through a long tunnel. Then a tanker that's too large to fit through slams into the tunnel at high speed, causing chaos. The important road is blocked, and the tunnel could collapse. The tanker is a traumatic incident, the damaged tunnel is PTSD.

Any direct attempt to pull the tanker out (i.e. face the traumatic memories directly) could cause the whole thing to collapse (i.e. experience the emotional trauma all over again). Psychological therapies for PTSD are like the workmen delicately reinforcing the tunnel, gradually cutting up and chipping away at the blockage, to re-open the road without further damage. There may be some unavoidable scars or long-term alterations, but normal functioning is essentially restored.

This metaphor is especially apt, because fixing the problem involves interacting with it directly. And so it is with emotions in the brain. Given the flexible, versatile, and extensively inter-connected way the brain is set up, the parts of the brain that process emotions are also those that produce them. The brain can't *totally* avoid experiencing the emotion it's processing, even if it's not a pleasant one, any more than someone learning to drive can't avoid getting into a car, even if they're claustrophobic.

The upshot of this is that, thanks to the flexible, adaptable nature of the brain, experiencing negative emotions helps you process them better.[92] Your brain gets more practice at dealing with them. Experiencing such emotions in non-disruptive, non-traumatic ways is better still. That's why art forms and entertainments that induce these (supposedly) negative emotions can prove so enjoyable, so helpful.[93] It's not that sad music makes you sad; it's that it's safe, cost-free sadness, without the pain or loss that usually produces this emotion. The brain gets all the benefits, none of the costs.

That's why sad music counterintuitively makes us feel *better*,[94] why angry music like heavy metal makes us *calmer*.[95] Besides the overall catharsis of experiencing emotions without any inherent risk, these emotional entertainments are like a short burst of therapy, boosting the brain's emotional abilities and resilience.

It also explains why teens seek out sad or angry music, or other negative emotional experiences, more than older age groups.[96] The still-developing adolescent brain hasn't 'figured out' how to process strong emotions yet, so any opportunity to experience the negative emotions they're constantly bombarded with, in a safe, consequence-free context, will be seriously appealing. And useful.

That's another key word: context. Do those who embrace sad music actively enjoy a relationship breakdown? Would horror fans be thrilled to encounter a genuine serial killer wielding a blood-soaked machete? Would a submissive BDSM enthusiast experience pleasure if a stranger kicked their door in and started whipping them? The answer to each, I'd wager, is a resounding no.

The emotion we experience, the effect it has on us, and how we process it, has much to do with the context in which it's experienced.[97] This reveals that our cognition, which recognises and determines what's going on around us, has another prominent role in the emotions we experience. The rational part of our brain is saying, 'This situation is safe, there is nothing to worry about here, you can put this book down or switch the TV off at any time, so it's OK to embrace this emotional stimulation'. And so, we do.

Once again, the division between emotions and cognition is nowhere near as clearly defined as many assume. Our emotions play a key role in how we think and rationalise, and our cognition plays a key role in what emotions we experience, and why. Another point in favour of the constructivist school of thought, perhaps?

So, if you enjoy sad movies or books, or listen to furious heavy metal, then don't let anyone tell you it's weird, or bad for you. You're doing your brain a service with your enthusiasms. It's like you're taking your brain to the gym, only with more crying.

Or less, if your enthusiasm for physical exercise is on a par with my own.

But if there's one thing that's cropped up repeatedly up to this point, it's that it's actually very hard to separate emotions from cognition. Because of this, I eventually ended up asking myself . . . could they, *should* they, be separated *at all*?

Emotions and thinking – same difference?

Earlier, I discussed the science fiction cliché that anything able to suppress, remove, or otherwise go without emotions will be superior to us feeble humans, with all our unhelpful emotions clogging up our mental machinery.

I also confessed that, thanks to my own emotional reactions (or, in some cases, lack thereof) to my father falling ill, I was coming around to that viewpoint, thinking that it might be nice to turn my emotions off for a while, and operate on pure cognition. As time wore on and not much changed, this possibility became increasingly appealing.

However, as this chapter has revealed repeatedly, emotion and cognition are intertwined far more extensively than I, and presumably many others, ever assumed. So, the question is, *can* emotion and cognition be separated? Can we ever really dispense with our instinctive feelings and exist in a state of pure reason? And if we can, *should* we? Is such a thing actually a good idea, given how the brain works?

This isn't just a wild fantasy or idle speculation on my part. Scientific research is, in many ways, designed to limit or remove the influence of emotions, in those responsible for conducting it.

Experimental methods work hard to limit observer bias.[98] Say a group of scientists spends years developing a drug that helps people lose weight. All being well, eventually they'll need to test it on humans. If human subjects take the drug and lose weight, this proves it works. Rewards would follow: major career boosts, lucrative pharmaceutical deals, international praise and respect, all that.

But, if human subjects *don't* lose weight, it shows the drug doesn't work. The scientists were wrong and must start all over again. Years of work, money, and effort, all for nothing.

Obviously, the scientists would really want to avoid that negative outcome. So much so, they may be tempted to 'tip the balance', to tweak the experiment, to make positive results more likely.

If they made sure all subjects had just started diets and joined gyms, that would obviously increase the odds of weight loss at the end of the experiment. Or, if some subjects show no weight loss, they could be excluded from the results, for valid-sounding reasons. 'This one's diabetic, that one's too old, this other one has an underlying condition', and so on. Basically, there are many ways to run an experiment to increase the odds of getting desired results.

The problem is, that's *not* science. Those results would be pretty much useless. It would be like a teacher who only counted correct answers when marking tests; every pupil would score 100 per cent, and the data would make the teacher look very good. But the data would be wildly inaccurate. They haven't got a class of precocious geniuses; they've manipulated the numbers to make it look that way. If this teacher got promoted and put in charge of the whole school thanks to these amazing results, it'd be a disaster.

This applies to science too, arguably even more so. Conclusions based on flawed, skewed data, which are then applied to real life, can cause serious issues. Especially in areas like medicine, where lives are genuinely in the balance.

Scientists know all this. But scientists are also humans, with human brains, so their actions and thinking can be guided by emotions (fear of failure, desire for success, anger at a rival, etc.) as much as by reason and logic. This means that scientists running and observing an experiment can, knowingly or unknowingly,* influence the outcome, to produce one they want. They introduce observer bias.

That's why scientific methods include things like control groups, randomisation, blinding, and more.[99] These exist to prevent the scientists running experiments from acting on their more emotional leanings, and thus spoiling the research.[100] And if scientists do manage to publish flawed, self-serving research, but are later found out, they can be stripped of their titles, position, and worse.

This has some interesting consequences, though. Have you noticed that scientists (or anyone similarly 'intellectual') are often portrayed, in the mainstream, as severely intelligent people who nevertheless struggle (or refuse) to form meaningful interpersonal relationships? From Asimov's Susan Calvin to modern-day Sherlock, to Sheldon Cooper of *The Big Bang Theory*, extremely smart people who nonetheless find emotions 'confusing' pop up very often in the mainstream.

* Remember that, thanks to how the brain is wired, emotion can indeed produce motivation without involvement of the cognitive, conscious processes.

This stereotype is perhaps unsurprising when we consider that science itself is constantly trying to remove emotions from the process. However, we've seen many things now, in this chapter alone, which show that this isn't how things actually work. Via Dr Mack, Dr Blackmore, and prominent philosophers of the past, we've seen that, rather than suppressing or ignoring them, many of our greatest scientists and thinkers ended up as such *because* of their emotions.

Being emotionally motivated to interfere with an experiment is unhelpful, especially because so much time and effort typically go into them. Even modest examples can require extensive planning, funding, running, analysis, and so on. Real science is, for want of a better term, a *slog*. A single experiment can be a years-long, often tedious daily grind, with no guarantee that anything useful will result from it.*

From a purely objective perspective, the tangible, concrete rewards of science are somewhat limited, especially when you consider all the effort and study required, just to be allowed to do it at all.

This creates a bit of a paradox: if scientists were completely logical and objective, which many seemingly think they are/should be, they *wouldn't choose to be scientists*. Not when there are far easier, and more financially rewarding, career paths available.

Nonetheless, countless people do. Why? For the respect of their peers? An ambition to be the best at something? A desire to help people or improve the world? A powerful drive to prove their ideas and theories are correct? A fear of uncertainty, or leaving the big questions unanswered? Because they simply enjoy research, and discovering new things?

Clearly, pure logic and reason isn't enough. People seemingly become scientists, and put up with all the drawbacks this entails,

* Another thing I have extensive personal experience in.

because they're *emotionally invested in it*, in some way. Ultimately, scientists need emotion too. It's just that emotions can't be expressed in the workplace. The cognitive regions of our brain may take centre stage with science, but they cannot do it alone.

Is this another interaction that can go both ways? If our emotions make us more rational and analytical, i.e. by compelling us to become scientists, can our rational minds, our cognition, cause us to experience irrational emotions?

Yes indeed. A good example of this is stage fright. The scientific label is 'performance anxiety', but either way, it's very common. It's where the very prospect of doing something, *anything*, in front of an audience makes you experience serious fear and apprehension, sometimes to an extent that makes you physically sick.[101]

Initially, stage fright looks like a clear case of emotions mis-behaving and causing trouble. There's practically never any actual physical danger in doing something before an audience, however bad you are at it.* The audience may be unimpressed and judge you harshly, but that's basically it.

But no matter how much we consciously, rationally try to convince ourselves of this, our fearful emotions keep coming. It can even occur days, weeks, *months* before setting foot on stage. Just knowing a performance will happen can be enough to trigger a severe emotional reaction.

Frustratingly, serious stage fright can still happen even if you have a lot of experience on stage. High levels of performance anxiety are seen in professional musicians.[102] Many even take medi-cations, like beta blockers, to reduce the symptoms of stage fright, as it can genuinely interfere with their performance.

All this points to a flaw or failure in the emotional processes in

---

* Unless it's some sort of physical combat thing, like boxing, mixed martial arts, etc. But even then, the danger isn't from the audience.

our brains. We experience serious emotional reactions to dangers that aren't actually there, to the extent that our rational brain regions, and our cognition, struggle to regain control.

Why would cognition struggle so much here? Well, maybe our cognition is *responsible* for the mess that is stage fright. Cognition may be failing to rein in the bull in the china shop, but it's also the one that brought a large, easily angered beast into such an unsuitable location in the first place.

Some suggest stage fright arises from a miscommunication between the brain hemispheres, where rather than cooperating to do something effectively, they get in each other's way. At least one study reveals that if activity in the left hemisphere is reduced and the right is allowed to dominate, then performance improves markedly.[103]

This is consistent with the view that the left hemisphere handles 'big picture' stuff while the right handles finer detail.[104] When on stage, the left hemisphere of our brain would be more aware of the audience (the thing we're scared of), while the right's concerned with the task, our performance. Quieting the former but not the latter would, logically, be helpful here.

Others allude to the Yerkes–Dodson curve,[105] which shows that, up to a point, stress and anxiety can *enhance* performance. This is consistent with the enhanced focus and concentration that comes with negative emotional experiences, covered earlier. Therefore, some stress is *helpful* when performing. Performance anxiety could be a useful trait, as while it makes us fear failure and embarrassment, the enhancing effects of that fear on our performance mean such an outcome is less likely.

But beyond a certain point, stress overwhelms our ability to cope and function. Performance anxiety becomes debilitating, counterproductive, by spoiling our performance. Why's it *so* stressful?

Well, we humans are incredibly social. We've depended on

the support and kinship of our tribe or group for much of our evolutionary history. Consequently, our brains have evolved to be extremely wary of any situation which could lead to the disapproval of others. Our brains typically respond to social interactions very favourably,[106] but they're also very sensitive to them going wrong, or badly, and that affects us in strongly negative ways.[107]

Our brains usually walk a fine line between managing social approval and social rejection. However, some suggest that, in the brains of individuals with serious performance anxiety (or other social phobias), there's an imbalance, and the potential negative consequences of engaging with others outweigh the positives.[108] Every performance becomes nerve-wracking, like trying to tap dance through a sleeping lion's den.

Of course, not *everyone* gets performance anxiety. Some are more prone to it than others. Many personality traits have been linked to stage fright, like neuroticism, perfectionism, fear of losing control, and more.[109, 110] It can even seem relatively mundane, like skewed perception of your own speaking abilities.[111] There are also psychological issues like catastrophic thinking where someone will persistently think about a worst-case-scenario outcome, regardless of how unlikely or irrational it is. That's obviously going to boost the likelihood of stage fright.

These tendencies and traits arise from somewhere. While there are some genetic aspects of eventual personality types,[112] most point to an individual's developmental experiences as the key factor in eventual performance anxiety (or personality traits that predispose to it).[113]

For instance, attachment issues[114] come up a lot. The bond between your child self and your parent (or primary caregiver) is seriously important for your overall development. So, say you have a particularly aloof parent who rarely expresses approval. As a child, you may end up valuing their approval much more, and/or dread disapproval, because they so rarely give approval that disapproval feels like a serious failure.

This all happens when you're very young, your brain is still forming, learning how everything works. This childhood experience could therefore form the basis of your lifelong perception and understanding of approval, causing your adult self to instinctively assign an above-average importance to the approval of others, with a corresponding sensitivity to disapproval. And so, you get serious stage fright.

What should be clear by now is that our emotions are by no means solely responsible for stage fright. We wouldn't respond with fear and anxiety to the idea of being booed by an audience if our cognition hadn't decided it was a real, even likely, outcome. Stage fright often comes down to overthinking the situation. Backing this up are studies suggesting that performance anxiety can be reduced or alleviated if you think about it differently, by coaching yourself to recast arousal and tension as excitement.[115]

Overall, stage fright reveals that our supposedly rational brain processes can easily result in illogical, unhelpful emotional experiences.

This sort of thing, the extent to which our emotions and cognition have such drastic, fundamental impacts on each other, kept surprising me. I eventually realised that this was because my whole approach was rooted in the assumption that emotions and cognition are different, separate things. That they're distinct features of our brains and minds.

Only . . . what if they're not? What if our emotions and our cognition, our 'executive functions', are not like two different colleagues working in the same office, but instead are like an individual's arms and legs? Limbs with different properties, different abilities, but part of the same body.

There is compelling evidence to suggest this is indeed the case.

Consciousness, as we'd recognise it, may have evolved *out of* emotions.[116] In the distant past, primitive creatures felt certain ways (i.e. experienced emotions) about events relevant to their survival, but never 'thought' about them, in any tangible way. But over time, the increasingly complex processing of emotions in increasingly complex species produced cognition and thinking as we know it today. That's the theory, anyway.

It's like how humans evolved from more primitive primates. But then, you often get those who dispute evolution saying things like, 'If humans evolved from monkeys, why are there still monkeys?', and promptly taking a victory lap to celebrate their mastery of arguments, before anyone can point out that they're talking absolute guff.

Because they are. Humans didn't evolve from monkeys; humans and modern monkeys evolved from the same ancestor species, like how a pencil and a broom handle could be made of wood from the same tree. The existence of the two different wooden things doesn't mean there was never a tree.*

Admittedly, a shared evolutionary origin doesn't mean emotion and cognition are the same thing. However, evidence is mounting which suggests that, in the modern human brain, emotions and cognition aren't nearly so distinct as is assumed.

Some studies suggest that emotional experiences at an early age are an integral factor in the development of executive control.[117] As in, the act of processing and responding to emotional stimuli is what allows our brains to develop these crucial cognitive abilities. Thanks to having to process and handle emotions, our childhood

---

* Also, asking why there are 'still monkeys' if humans evolved from them is like saying, 'If adults grew from babies, why are there still babies?' That's not how evolution works. Also, if we've got adults denying evolution purely because monkeys exist, we need better adults.

brains are shaped and developed in ways that allow us to practise self-control, expectation, deduction of the 'If I do this, then that happens' variety, and more. These are the basic building blocks of our executive control, our cognition.

On the other hand, some studies flip this arrangement and suggest that conscious control and executive functions are essential for the proper development of emotions.[118] Much research has noted that the processes underlying emotion and *control* of emotions (i.e. conscious self-control, executive function) appear basically the same, in terms of neurological activity.[119] It further emphasises that the brain doesn't readily discriminate between emotion and cognition nearly as much as is assumed.

Models like the triune 'three distinct evolutionary layers' brain suggest there are clear boundaries between different parts of the brain, and that our conscious self is the result of their combined output. It suggests the brain is like three small children on each other's shoulders, in a long trench coat, trying to sneak into a cinema. It sounds ridiculous, but it works, because in this scenario everyone else is *also* three kids in a trench coat, so nobody thinks anything is amiss.

However, anatomical, physiological, and neuropsychological evidence has long ruled out such clear functional divisions in the brain. It's actually composed of many complex and widespread networks of different brain parts, often with the same parts working in different, distinct, ways. Brain regions are often pluripotent: one thing, with many possible functions.

Earlier, I described the circuit in the brain that's believed to give rise to our emotions and our individual ways of processing them.[120] The one that includes the prefrontal cortex regions, the amygdala, the hippocampus, the anterior cingulate cortex, and so on. We saw how widespread and multifunctional it is, and how different parts of it do so many things, like how the amygdala is a hub of emotional processing, with many roles and extensive connections, to both cognitive and emotional systems and networks.

This is similarly true for the anterior cingulate cortex, another part of the emotional circuit, and a brain region long associated with emotions.[121] It has a wide range of functions, from decision making to pain perception to social behaviour guidance. It's also integral in assigning emotions to stimuli, as well as determining how we respond to them.

Given the important and varied roles of the anterior cingulate cortex, it has copious connections to the rest of the brain and handles both emotional and cognitive information. Until recently, it was believed that the cingulate cortex kept these information streams separate, that some bits were responsible for conscious information, some for emotional, and there were clear separations between them.

But more recent evidence suggests that this isn't the case after all, and that areas presumed to be specifically for conscious processing have emotional roles, and vice versa.[122, 123] All this does suggest that emotion and cognition are actually more like two alternative expressions of the same thing, different limbs of the same body. Or maybe it's even less rigid than that? Maybe it's like a river that forks into two separate channels before it reaches the sea. One channel is emotion, the other is cognition. Same water, same origins, but different destinations?

At the root of all this uncertainty is the perennial question that comes up in this area of research. Namely, when we refer to an emotion, *what exactly are we talking about*? Does the reaction to it count? The motivation it produces? Our perception of it? The effects it has on our thinking? An emotional experience includes all these things, and more. Are they a valid component of an emotion? If not, why not? And how do you disentangle the 'pure' emotional processes in the brain from those that are more incidental? That seems to be beyond us at present.

And maybe . . . that's good? Given everything I'd uncovered so far, it seems increasingly the case that trying to find the 'true essence' of emotions in the brain is like trying to find the humour in a joke by removing all those unnecessary words. Can't be done. Not how it works.

It's safe to say that, at this point, the notion of separating my emotions from my rational thinking and just relying on the latter seemed incredibly unwise, let alone impractical.

I was reminded of something else that Dr Firth-Godbehere brought up when he and I spoke.

Given the nature of his work, he flagged up Lieutenant Commander Data, a character from *Star Trek: The Next Generation*. Data, portrayed by actor Brent Spiner, is an advanced android who doesn't have emotions. He's manifestly stronger, smarter, faster, and more capable than any human, but he's still regularly trying to *be* more human, specifically regarding his lack of emotion.

Data was a very popular and iconic character for the franchise. But, based on what we know now, about emotions and their role in our cognitive abilities, a real-life Data, a self-aware intelligent machine with no emotions, would be very different. As Dr Firth-Godbehere explained memorably:

Technically, if you asked Data to, say, pick a flavour of ice cream, he wouldn't be able to. How could he?

Data's mind is, presumably, based on pure logic and reason . . . but there's no logical basis to prefer one ice cream flavour over another. Especially when you're a machine and have no need for sustenance. Without resorting to making decisions entirely at random, something that computers and software have always found tricky,[124] Data would have no cause to pick one type of ice cream over another.

It's fun to imagine the super-advanced android, stood at the counter in an ice-cream parlour and staring at the menu, as frozen as the confection on offer, while an increasingly irate queue builds up behind him. But the implications are rather more profound.

If there's one thing we've learned in this chapter, it's that, for a great deal of what we perceive and how we perceive it, what we're motivated to do, how we think about and assess information – for all this and more, our emotions play a significant, often crucial role.

So, while much of sci-fi suggests that removing the influence of emotions would make us better, smarter, more mentally capable and ruthless, in reality it would leave us cognitively crippled, unable to think or do much of anything at all.

Even if you could completely separate, or distinguish between, the cognitive and emotional processes in our brains, suppressing or removing your emotions wouldn't be like clearing an obstruction. It wouldn't be like taking out an inflamed appendix, or removing a paper jam from an expensive photocopier.

No, it would be more like removing all the mortar from your house and just leaving the bricks.* It wouldn't improve anything, it'd just bring the whole thing crashing down, leaving devastation and rubble.

And that's why the idea of removing my emotions lost all its appeal. I may have been ignorant of them, and I may have been constantly confused, distracted, exasperated, or frustrated with them at this particularly difficult time in my life. But it became abundantly clear that my emotions aren't holding me back; they *are* me. They're so fundamentally ingrained in my brain that they're an integral part of my mind, my identity, my ability to exist as a thinking being. And the same goes for everyone else.

Even if I had been having issues with them, that doesn't mean

* How exactly you'd do such a thing is a whole other problem.

my emotions should be discarded. This would be like hacking your leg off because of a splinter in your toe. Only worse, because if you hack your leg off, you're still able to think about what a stupid idea that was afterwards.

No, wherever I ended up on this weird journey my emotions were taking me on, they were here to stay. I was never going to get rid of them.

Not that it was ever even an option, admittedly. I'm not a fictional scientist, I'm a real one.

I will admit, though, that, at this point, all this information I was uncovering about emotions was becoming a lot to take in. However, if there's one thing that emotions have a profound impact on, it's memory. Something I was about to discover first-hand, in the most brutal of circumstances . . .

3

Emotional Memories

In late April 2020, despite the best efforts of the medical profession, my father died from COVID-19.

Did I cry at that point? Indeed. The logjam holding back my emotions crumbled, and . . . stuff came out. But it was often at odd times, and in odd ways. When I learned he was gone, I was weirdly numb for several hours, before spontaneously collapsing into a mess that evening.

This continued for several days. Anything could set me off: seeing the colourful shirt he gave me for Christmas; the smell of aftershave (he always doused himself in it); mention of a birthday that he would no longer be part of. Anything that caused me to remember my father, and that he was no longer there, hit me hard. And it hurt.

However, as ever, the analytical parts of my brain were still whirring away, which led me to notice something odd. People who wanted to help me through my grief regularly advised me to 'focus on the good memories'. It made sense, but there was a problem: the good memories of my father were suddenly painful. They were now infused with a potent sense of loss.

We're used to thinking of memories as fixed, unchanging records of our experiences and knowledge, like the files in a computer, or the words in an old diary. But that's not how the brain does things; our memories are far more flexible, more changeable, than that.

I should know: my PhD was in how complex memories are formed and retrieved in the brain.[1] And if there's one thing that we know plays a vitally important role in how memory operates, it's emotion.

Yet I'd never really dwelt on exactly *how* emotions influence memory, in the scientific sense. It's something I (and many colleagues) took for granted, devoting our efforts instead to the more cognitive or neurological aspects of the brain's memory system. I now realise this is the neuroscientific equivalent of praising a jockey for winning a race; we all do it, but it's somewhat unfair on the horse who actually did all the work.

So, I resolved to find out exactly how, and why, emotions have such a big part to play in our memory. Maybe in doing so, I could figure out what to do about my own conflicted, confusing memories.

Or at least I'd be distracted from them for a few hours. At the time, that seemed as good an outcome as any.

Remembering the good times

As I write this sentence, my father's death happened mere weeks ago. It's *very* clear in my memory. Distressingly so. But I have to wonder: if I revisit this part of the book, months from now, will the memories of my father's passing have faded, become vaguer, less painful?

I doubt it. I confidently predict that these memories will remain crystal clear, possibly forever, because they are powerfully emotional. And emotional memories are inevitably more robust and enduring than 'neutral' ones.[2]

This happens all the time. We've all experienced the feeling of spending countless hours preparing and studying for a big exam or work presentation, only to find that, once it's all over, we're only dimly aware of the information we put so much time and effort into learning.

Admittedly, the human brain is better at retaining information associated with an unfinished task, but quickly forgets it when the task is completed. It's known as the 'Zeigarnik effect',[3] first studied in the context of restaurant waiting staff, who regularly remember complicated food orders for large groups while serving them, but promptly forget them once those customers are dealt with.

Also, sitting exams or giving presentations are rare events for the typical human, and how often are your employer's third-quarter-sales estimates brought up in everyday life? That's a very niche pub quiz. If such memories aren't activated, they can atrophy, like an underutilised muscle.

However, we also struggle to recall such abstract information because it has no *emotional* element. Our brains have more difficulty committing such things to memory. To understand why, it's important to recognise that human memory is complex, works in many different ways, and takes various forms.[4]

Some memory happens without us knowing. This is implicit memory. It's just like riding a bike. Literally: getting on a bike and knowing how to ride, without thinking about it, is a form of implicit memory, labelled procedural or 'muscle' memory.

Other implicit memory types include habits, like brushing your teeth in the same pattern every time, and associative or conditioned responses, like reflexively rejecting or feeling queasy when offered a food that once made you sick. These all require remembering things, but we aren't aware that we're doing so. By definition, the conscious, cognitive parts of our brains aren't especially involved in implicit memories.

Procedural memories, for motor skills like bike riding, depend heavily on the cerebellum,[5] the wrinkly bulge emerging from the bottom of our brain, just behind the brainstem. Conditioning and associative learning are handled by areas like the striatum,[6] a prominent part of the basal ganglia.[*] These regions can, demonstrably,

---

[*] A cluster of fundamental neurological regions, located deep in the centre of the brain, with many vital functions.

access important memories without getting our cognition involved. The striatum and cerebellum are also both known to have multiple important roles in emotion,[7,8] suggesting that emotion plays a role in implicit memories. Logically, if you unthinkingly recoil from a disgusting food, you must have experienced disgust at some point.

The more familiar type of memory, the type of memory you are aware of accessing and recalling, is explicit memory. Explicit memories are formed by the hippocampus[9] and retrieved, remembered, via the prefrontal cortex.[10]

Explicit memory can be divided into two types: semantic and episodic. Semantic memory means abstract information, without context. Or, in plain English, it's the stuff you know, but don't necessarily know *how* you know it. For instance, I know that Montevideo is the capital of Uruguay, but I couldn't tell you where or when I learned that fact; ergo, it's a semantic memory.

Episodic (or 'autobiographical') memories are of first-hand experiences in our lives, and include information about the context in which the memory was formed. And, as you may have noticed, our most enduring memories are typically those of episodes that were the most joyful, heart-breaking, embarrassing, enraging, terrifying, or involved any other type of powerful emotional experience.

This happens because emotions directly enhance the brain's memory system. Long-term explicit memories are formed by the hippocampus.[11] Every specific experience we have is made up of distinct elements: the sensory feedback our brain is receiving at that particular time, and things going on inside us, such as our mood, physical comfort, how tired/hot/cold we are, etc. All this is relayed to the hippocampus, which creates a memory for this particular combination of elements.

A memory is stored in the brain as a specific collection of synapses, connections between neurons.[12] Synapses are the fundamental components of memory; they are to memories what the zeros and ones on a hard drive are to software.* Accordingly, the hippocampus is one of the few parts of the brain where new neurons (brain cells) are known to be created during adulthood.[13] They're necessary for creating new synapses, i.e. memories.

However, consider just how many elements there are for your senses to be aware of in a single experience: every item visible in your surroundings; every sound you hear; every smell; the people you're with; their expressions and body language; lighting; time of day; the twinge in your leg; and so on. All this, and more, can be relayed to the brain, in a single second, as you stand in a boring supermarket queue.

The hippocampus cannot turn every element of every experience into a memory. Even our mighty human brains don't have that sort of capacity or processing power. Even if they did, how much of what you actually experience, moment by moment, ends up being important later? Consequently, the hippocampus, and associated systems, prioritise certain experiences above others.

While the information needed to pass our exam or deliver our presentation is objectively, *cognitively*, important to us, that's not how it works. The memory-formation system predates our more sophisticated cognitive abilities. So, our memory system typically decides that the more viscerally stimulating or significant an experience is, the more important it is to commit it to memory. And this stimulation/significance is heavily shaped by emotions.

This happens via that go-to neurological region for emotional processing, the amygdala. The amygdala is located right next to the hippocampus, and the interaction of these two regions is a

---

* The combination of synapses that forms a specific memory is known as an engram. Given the limits of our technology and the baffling complexity of the brain, the engram is still technically a theoretical concept, although modern developments seem close to proving it as a practical reality.

well-known and important part of the memory-formation process, specifically when it comes to emotional memories.[14] As we've seen, the amygdala is extensively linked to countless other parts of the brain. For example, when you see someone's facial expression, you quickly recognise the emotion it represents, even feeling it yourself, to an extent. This is caused by rapid and direct links between the amygdala and the visual cortex region that perceives faces.[15]

If the information entering the brain is like raw materials on a production line, the amygdala is a foreman stood right where they hit the conveyer belts, tagging any emotional material as 'high priority', and speeding it along to the relevant destination, with instructions about what to do with it.

That's why, as we saw previously, the emotional quality of an experience influences how focussed we are on it. Studies reveal that people can find images of snakes or spiders much faster than emotionally neutral images, in exactly the same setup,[16] suggesting our attention is directed towards anything that presents a threat, i.e. causes fear, before we're consciously aware of it. Other emotions seemingly have similar effects.[17]

Interestingly, evidence suggests that happy memories contain much more peripheral information – details not specifically relevant to the main event – than unhappy ones. The background music playing when you proposed to your fiancé; the colour of the waiter's hair at your surprise birthday party; and so on. Negative emotional memories, by contrast, lack such external details.[18]

This is consistent with what we've learned about positive emotions widening our cognitive scope, while negative ones narrow it. It therefore makes sense that the memories for these emotional experiences would reflect the sort of information available to the brain when the memory was formed.

It seems the amygdala inserts the emotional component of memories. Damage or disruption to the amygdala can reduce or remove the hippocampus's ability to form emotional memories.[19] The hippocampus can still form memories for emotional events *without* the amygdala, but those memories will be less significant.

For example, have you ever got very sad or angry about something when drunk, only to find that you later can't recall why you were so upset, and are baffled by your own reactions? Alcohol disrupts memory formation,[20] so maybe such intoxication interferes with amygdala–hippocampus communication, hindering the integration of the emotional aspect of the experience into your memory. Your hippocampus records the details, and the fact that you *had* an emotional reaction, but the actual emotion itself is lost.

The amygdala also directly enhances memory formation by increasing relevant activity in the hippocampus and other memory-processing areas. This is the 'modulation hypothesis', because the amygdala modulates (i.e. changes) what's going on in the hippocampus (and other areas) when significant emotions are felt.[21]

Basically, when experiencing powerful emotions, the amygdala is like a technician turning up all the dials on a sound desk (i.e. the hippocampus), so that everything is amplified. Memories formed at this time are therefore more potent, more significant, and easier to recall.

This relationship can also work in reverse, i.e. the hippocampus and associated memory system can influence the emotions we experience, by acting on the amygdala. Put simply, our memories can often dictate our emotions.

For instance, are you afraid of flying? Did you feel apprehension and dread the first time you set foot on an aeroplane? This means your amygdala was firing wildly, in response to the danger it was recognising.

Why would it do that, though? You'd never been on an aeroplane before, so your subconscious emotional processes should have no reason to produce a fear response. However, the brain is capable of learning about and understanding aeroplanes, and what they involve, without ever going near one. Therefore, we can be afraid of the *idea* of something, without having to experience it first-hand.

To put it another way, your abstract memories about aeroplanes and what they mean are enough to trigger powerful emotions. The amygdala is reacting just as rapidly as ever, but what it's producing a fear reaction to in this case comes from the memory, not just the senses. So, our memories can influence the emotions we feel, just as emotions influence the memories we retain.

As with stage fright, there are many complex factors under-pinning a fear of flying.[22] However, scientists have demonstrated this phenomenon in a far more straightforward way. In one study, subjects were told that seeing a blue square meant they'd get a shock, and subsequently displayed a clear fear reaction on seeing said blue squares.[23] The *cognitive* representation of danger, based entirely on memory, triggered an emotional response. The neuro-logical association between emotion and memory is clearly a two-way street.

Once formed, memories need to be effectively stored, integrated into the existing vast networks of memory and information so they end up in the right place. It's not that newly formed memories can't be used immediately, because of course they can. But making them as robust, enduring, and effective as possible takes time. This process is known as consolidation.[24]

It's like a truckload of new books being delivered to a library: the books are readable as soon as they arrive, but to be useful to the library they need to be catalogued, filed, and put on the correct shelves. The brain does similar with new memories.

One argument about why consolidation takes so long, at least initially, revolves around emotion. The initial stages of consoli-dation, where new memories slowly move out of the hippocampus to wherever they're needed, proceed gradually. It's argued that this slow pace is an evolved feature, because the emotional experience is an important aspect of any significant memory, but often occurs *after* the event.[25]

When we feel angry, embarrassed, guilty, or pleased, it often takes our brain a few seconds to realise that this is warranted emotional reaction. You take the last slice of pizza, but then your partner says they wanted it, so you feel guilt. You're in an office meeting and a co-worker says something to the boss that, minutes later, you realise was a direct criticism of you and your work, so you get angry. In both cases, the emotion occurs *after* the event itself.*

This occurs on a chemical level too: stress hormones, like cortisol, enter the bloodstream and have many effects on the brain and memory systems, but only *after* the event that caused the stress.[26]

However, when you recall these instances from memory, you'll remember the emotions you felt, even though they were experienced later. You don't have two separate memories, one for your co-worker saying something negative about you, and another where you got angry, seemingly apropos of nothing.

Memory consolidation occurring slowly allows time to add an emotional reaction to the still-fresh memory, before it's 'set'.[27] It's like the other elements of a memory are in a lift, waiting to go down to the factory floor, but emotion is still lumbering down the corridor. Because emotion is so important for memory, the brain holds the lift door open, so nobody can leave until emotion joins them.

Here's the thing, though: it's clearly not just the immediate stages of memory consolidation where emotions change things. I said that the loss of my father meant all my happy memories

---

* This doesn't change how quickly the amygdala responds. It's as fast as ever, but in these situations, it doesn't realise there is anything to react emotionally to, until after the fact.

of him were tinged with sadness now. These memories stretch back over the *four decades* of my life. So, even old memories, which have been completely consolidated for many years, can be altered by a later emotional experience. And this has several profound consequences.

Say you're at a party and are introduced to a friend of a friend. You exchange hellos and pleasantries, but quickly move on to speak to someone you actually know. You may never speak to, or even think about, this friend of a friend again. What memory exists of this encounter will be deemed unimportant. At best, it'll just sit there, figuratively gathering dust in the recesses of your brain.

But then, you're watching TV one day and they pop up on the news because they committed a series of grisly murders that took place in, for example, an aquarium. Suddenly, that original memory of them is *very* important. Before, you'd probably struggle to remember even meeting them. But now, thanks to the emotionally-charged new information, you vividly recall the time you met the infamous 'Sea Life Slasher', and will probably never forget it.

This phenomenon is 'retroactive memory enhancement'. Recent studies reveal it happens in human memory quite readily.[28] It basically means that an emotional experience *now* can enhance a memory from a long time ago, even if that memory was largely insignificant and little used before the later emotional experience.

Doesn't this suggest, though, that we actually *do* remember everything we experience, even if it is mundane and irrelevant, despite me saying earlier that this is impossible? Well, not quite.

The neurological processes that lead to forgetting memories are complex and varied. Sometimes newer memories interfere with or overrule old memories, so the brain defaults to the newer one.[29] Sometimes, newer neurons supporting newly formed memories alter the hippocampal network, so existing memories, particularly those still dependent on the hippocampus for access, are disrupted, and lost.[30]

And in recent years, scientists have identified 'intrinsic forgetting', whereby specialised brain cells actively remove memories that aren't used.[31] It's seemingly an ongoing process, and memory consolidation is regularly working against it, like someone constantly rebuilding and reinforcing sandcastles on the beach as the tide comes in.

Counterintuitive as it may be, forgetting seems to be the default state of the brain's memory system. As the hippocampus is constantly logging elements of every experience, the brain's storage capacity would quickly be used up if it was all retained forever. Instead, memories we don't need, or use, are constantly cleared away.

Similarly, not every experience will lead to a brand new memory being formed. Memory is based on *connections* in the brain: our brains have access to all our existing memory traces, and can incorporate these into new memories being formed, thus saving energy, space, and resources. So, we have a specific dedicated memory of, say, our spouse stored in our brain, and that memory is then linked to all the memories of the experiences you have together. A much more efficient system than creating wholly new memories of your spouse every time you encounter them.

This point about connections being the basis for memory is a crucial one, because if a specific memory is a specific combination of connections, then there's no reason why more connections couldn't be added to this memory later.

For instance, my father often bought me clothes as a Christmas present, so I have a lot of shirts from him. I had no problem with them before, but since his passing, it feels weird, melancholy even, to wear them. It's not that I see these shirts and think, 'These are sad shirts now'. No, the shirts make me sad, because they remind me of Dad. My father's death inserted a profound element of sadness into all my memories connected to him, including those concerning the origins of the shirts hanging up in my cupboard. The inanimate, unchanged garments now trigger an emotional response, because they're connected to memories of a specific person.

Studies suggest this is largely why people have keepsakes and heirlooms.[32] It's not that all those fridge magnets or snow globes necessarily make us happy in their own right. Rather, they help us remember the people, or events, that they're connected to in our memories.

This can be particularly important in old age, when we have more memories but less to look forward to. Our grandparents' pictures and knickknacks may look like pointless clutter to us, but research reveals that a lack of keepsakes or memorabilia in older people is linked to low mood or depressive tendencies.[33]

This process can be negative, too, as anyone who's gone through a bad breakup and thrown out (or even *burnt*) everything associated with their ex would attest. It's the same principle: you don't hate the inanimate objects with destructive intensity, but they make you remember a person you now *do* hate. And if the objects 'represent' that person in your memory, actively destroying them could provide some helpful catharsis for the pent-up anger you feel, in a way that doesn't hurt anyone.*

There are downsides to this, though. Suppressing/avoiding unpleasant memories seems to inhibit consolidation, which impairs recall.[34] To put it another way: by not engaging with the memory, it'll be harder to remember it later.

This may sound like a positive, especially if it's a particularly bad

---

* It's not great for the objects themselves, but still better than setting fire to your actual ex.

breakup. However, there's a reason that emotional memories are so potent: they're *useful*. It's not nice to remember a bad breakup with intense clarity, but what if you later get romantically involved with someone new, who has many similar qualities to your ex? People often have a 'type', after all. Remembering the anguish experienced the last time may prevent you from making similar mistakes or unhelpful decisions now. In this way, suppressing such emotional memories can be like forgetting you're allergic to a certain food; it's not a *nice* memory, but it's certainly a helpful one.

On the other hand, our brains sometimes push this process too far. Vivid memories of a bad breakup can make you paranoid and suspicious about *any* future romantic involvement, which is self-sabotaging, preventing you from moving on and finding happiness. Similarly, constantly triggering memories of a deceased loved one can keep the grief very potent, hindering your ability to cope and find acceptance, to 'move on'.[35]

Basically, sometimes it's bad to suppress emotional memories, and sometimes it's good. How can you tell which one is which? If you figure that out, do let everyone know.

Overall, it's increasingly clear that, due to the connective, plastic way the brain works, existing memories, even important ones, can be changed, or updated, due to new experiences.[36] It's like adding a digit in your existing password when it's time to update it.* The password has only undergone a minor change, but that change is still vital, and the older version of the password no longer works.

It makes evolutionary sense: the world around us is constantly changing, so tweaking the established memories we use regularly is a helpful trait, otherwise we'd be constantly basing our actions and decisions on outdated information. And, as we've seen, our

---

* I know a lot of people do this, but I'm told it's not good for proper web security, so I'm *definitely* not endorsing it. It's just an analogy.

memory systems are heavily influenced by emotions. So, an emotional experience has a lot of scope to alter your memories.

What does that mean for me, though? Did the negative emotions I experienced after he died diffuse through all the memories of my father, like a drop of black ink through clear water? And does the enduring nature of emotional memories mean they are changed *forever*? Whatever happened to 'Time heals all wounds'? Doesn't the influence of emotion over our memories mean that this notion is nonsense?

Luckily, no. It seems that, for once, the brain has done us a favour, and devised a system to prevent this. It's called the fading affect bias.[37]

Negative emotions are typically more potent and impactful than positive ones.[38] Most people will recognise this; the most joyful experience of your life still won't affect you as profoundly, and as enduringly, as the most painful. And any performer will tell you that, when facing an audience, it's the one miserable face out of hundreds of smiling ones that you remember.

This could stem from how our negative emotions are tied in with threat detection. We would logically focus more, instinctively, on things which present 'a danger' to us. Or, it could be that negative emotional experiences are more diverse; you can be angry, disgusted, fearful, guilty, and so on. By contrast, positive emotions are mostly variations on happiness. Therefore, potentially more parts of the brain are active when we experience negative emotions, making them more prominent as a result.[39]

Thankfully, even if negative emotional memories are more potent, the fading affect bias means they don't last as long as positive ones. The negative emotional qualities of memories fade relatively quickly, while positive ones linger.[40]

It's not that we forget the emotional incidents; it's more that the ability of those memories to induce that emotion diminishes over time. Eventually, the memory of an injustice will make us think, 'I was angry about that', whereas before it was, 'I *am* angry about that'. The same applies to other emotional experiences: we remember what we felt, but recalling it doesn't make us feel that way anymore.

Positive emotional memories are different, tending to produce positive emotions in us for much longer. Unless they were particularly traumatic, unpleasant incidents from our childhood usually fade, while happy memories still make us smile decades later. This explains why looking back with rose-tinted glasses is so common. Even if their past wasn't that good, people still tend to remember it fondly, because only the good parts of it still resonate in their memories.

Some evidence suggests this is another evolved trait, to maintain good wellbeing and a sense of self-worth, to keep us motivated, and so on. Jettisoning bad memories and keeping the good would logically help us feel better about ourselves, long term.

It's also worth mentioning that the fading affect bias is a lot less pronounced, or absent altogether, in people experiencing dysphoria,[41] a state of unease or dissatisfaction common in depression and similar mood disorders. Such conditions can be very stubborn and enduring, which is unsurprising when you consider that, for those experiencing them, one of the brain's default mechanisms for getting rid of bad or negative emotions is compromised.

What can't be denied, though, is that the way our brains make memories is more flexible and complex than we assume, and our emotions are a very big part of that. But even though our emotions can clearly change our memories in profoundly negative ways, this effect doesn't last forever, and over time, for better or worse, the bad stuff fades and the good stuff remains.

So, when people are telling me that I should 'concentrate on the happy memories' of my father, it's difficult, because, thanks to my own grief and the way emotions alter memories, they aren't really 'good' memories anymore.

But, thanks to how the brain works, soon enough, they will be again. After all, time heals all wounds, right?

Having said that, if recent experiences are anything to go by, it'll probably be a while longer again before I can smell Dad's signature aftershave without being emotionally overwhelmed by it. And, as it turns out, there are good reasons for that.

The scent of emotion

Recently, while out for an evening stroll, I encountered the smell of cigarette smoke and experienced a brief burst of reassurance and happiness.

This was weird, because I'm not a smoker. Never have been. I've always found it flat out unpleasant. Even when many friends took up smoking during my teens, it never tempted me.

It's not that I've never tried it, because I have. I wanted to find out if there was some positive aspect of smoking I might have overlooked. I am a scientist, after all – experiments go with the territory.

Also, I was a drunk student at the time.

I recall I did actually experience a vague pleasurable buzz when I tried my first cigarette, but this was immediately overshadowed by the violent reaction from my lungs, which strongly objected to the whole endeavour. Add to that how my mouth felt like an incontinent badger had been hibernating in it, and it's safe to say the appeal of smoking eluded me still.

I even tried again later, completely sober, just to be thorough. Same reaction. So, even overlooking all the known health risks, smoking clearly isn't for me.

And yet, despite my overwhelmingly negative experiences of smoking, when I recently smelled cigarette smoke, I felt good, reassured, and a strange form of contentment. Basically, I experienced a positive emotional reaction. Why?

I concluded that it must be my memory system that was behind the phenomenon. In the wake of his death, I've obviously been thinking a lot about my father, my family, and my childhood. I grew up in a lively pub in a working-class Welsh mining valley in the 1980s, where Dad was the landlord. The UK smoking ban was decades away, so cigarette smoke was the background odour of much of my early life. Therefore, despite my own unpleasant experiences with it, the smell of smoking remains linked with happier carefree times, with more positive emotional memories.

However, then my neuroscience training kicked in and made me realise that this explanation still didn't add up. The thing is, I found smoking viscerally disgusting when I tried it, and, thanks to how our brains work, when something disgusts you, no matter what's gone before, that's usually the dominant association in your memory.[42] If you loved halloumi cheese and ate it all the time, you'd end up with many positive memories of halloumi. But if you eat one spoiled piece that makes you sick, *that's* the memory your brain will cling to, regardless of what's gone before. It's a very powerful process.[43]

Except, it seems, when it comes to smell. The smell of cigarette smoke, for me, remains linked to positive emotional memories, despite my more recent unpleasant experiences. Why does smell subvert the norm?

In truth, many people have observed, over the years, that certain smells trigger emotional responses and memories, far more potently than most other sensory stimuli.[44] So, does our sense of smell have a special relationship, within our brain, with the workings of memory and emotion?

The answer is yes, very much so.

We don't usually give much thought to our sense of smell, we humans. Our nasal prowess falls far short of that of our animal friends like dogs and cats. We rely far more on vision[45] and hearing.[46] But despite this, our sense of smell is affecting us in very potent ways that we typically aren't aware of.

Our sense of smell is produced by our olfactory system. Odour molecules in the air enter the upper chambers of our nose, the nasal cavity. This space is lined with the olfactory epithelium, a tissue layer containing many olfactory receptors, embedded in olfactory neurons, which detect and recognise odour particles, and send the relevant signals to the brain.[47] Our olfactory epithelium is to smell what our tongue is to taste.

The olfactory epithelium is coated in a constantly replenished layer of thick mucus, into which odour particles dissolve, to enhance detection by the olfactory receptors. Signals from the olfactory receptors are transmitted to the olfactory bulb, the brain region that processes information about smell.[48] This, like most parts of the brain, has many complex subdivisions, and many connections to other neurological regions and networks. But delve a little deeper, and things take a turn for the surprising.

The genes that code for olfactory receptors take up 3 per cent of our genome.[49] Smell is also believed to be the *first ever* sense to have evolved.[50] That may seem remarkable when you consider how much more detailed senses like vision and hearing are, but it makes evolutionary sense when you consider that Earth's earliest life forms were made up of basic cells, little more than complex bags of chemicals, in a chemically rich environment, like primordial soup or ancient seas.*

From that perspective, rather than light, sound, heat, or pressure, the most important thing for the earliest life forms to be able to sense, in terms of survival, would be a chemical change in their surroundings. And what is smell but the ability to detect chemicals

---

* It's uncertain exactly where life first emerged from on Earth. Nobody around today was there at the time.

in the environment around us? We've come far since the primordial ooze. But in some ways, not *that* far.

The fundamental importance of smell produces interesting aspects of how it works in the brain. The neurological regions that handle our other primary senses (hearing, vision, etc.) are found in the neocortex, the top layer of the brain.* The olfactory cortex, by contrast, resides in the limbic regions, lower and more central in the brain, nestled right amongst the areas responsible for emotion and memory.

Indeed, when it was first identified, the hippocampus was assumed to be part of the olfactory system, so close and overlapping was it with the regions known to be involved in our sense of smell. Its crucial role in memory was established later.

This isn't a coincidence; the hippocampal and olfactory systems aren't completely different things that just happened to end up beside each other, like a heavy metal band that unknowingly moves in next door to an uptight vicar. No, evidence suggests the olfactory system and the hippocampus evolved *together*; they influenced each other's development because they're fundamentally *linked*.

Why would smell and memory be so closely bonded? Well, another key function, perhaps the original key function, of the hippocampus is navigation.[51] Countless studies have shown the hippocampus as essential for our ability to navigate through our surrounding environment, like the famous study which demonstrated that experienced London taxi drivers, who've spent years memorising how to navigate around the large complex city, have larger than average hippocampi.[52]

To navigate, you need to know where you are and where you've been. The hippocampus records locations of the useful landmarks

---

* Although they are all, as ever, extensively connected to lower brain regions.

around us, meaning our brain can utilise this information to track the change in said landmarks, relative to our own location, and build up a cognitive map of our surroundings,[53] allowing us to work out where we are and where we're going.

Essentially, the hippocampus supports navigation because it recognises and stores specific arrangements of sensory elements for later use, exactly as it does for memories. The only difference is that memory formation is not limited to spatial information. In fact, our entire memory system could have arisen from our primitive ancestors' need to know where they were going and where they'd been.

Where does smell come into this? Well, for a long time, smell was the dominant, maybe *only*, sense that living organisms had. However, the ability to sense things isn't much use if you can't do anything with that information. So, life forms also needed to be able to figure out, subjectively, where things are, and to move towards or away from them, depending on whether they're good or bad. Basically, as soon as we were able to sense our external environment, we'd need to use that information to navigate around said environment.

As such, the olfactory system and the hippocampus have evolved in tandem for aeons, arguably shaping the whole structure and layout of modern brains,[54] with the other, now dominant senses being added to the network later. Given this, it's only natural that smell and memory would overlap in many ways.*

Smell is also our oldest sense in developmental terms; we acquire it in utero,[55] and it is believed by many to play a fundamental early role in cognitive development.[56] Given this developmental head start over the other senses, smell is bound to feature more

---

* The olfactory system is also constantly producing new neurons, as existing ones get rapidly degraded by exposure to the 'outside world' (i.e. the nasal cavity). Another thing it has in common with the hippocampus.

prominently in our early life, and therefore our earliest memories.

Studies back this up. While memories triggered by visual or audio cues peak in our teens, memories triggered by smells stretch back about a decade further, mostly stemming from between the ages of six and ten.[57] Put simply, the memories triggered by smells can be a lot older than those triggered by other sensory stimuli. The claim that certain smells can trigger vivid memories of our early days is scientifically supported.

Also, earlier smell memories do indeed seem to overrule later ones, in a way that can't be said for the other senses. For some reason, the brain privileges the *first* association it makes with a smell,[58] and later experiences that contradict it are less influential. So, that's why I still think fondly of cigarette smoke and associate it with my childhood, despite my very negative adult reaction to it.

Memory for smell is also typically more vivid than memory for other senses. This is probably also the result of the olfactory cortex's special relationship with the hippocampus. The other primary senses are linked to the hippocampal memory system via the thalamus,[59] a vital region deep in the centre of the brain, which relays information from certain parts of the brain to those where it's needed. This includes sending sensory information from where it's produced to the hippocampus, so it can be turned into memories.

This isn't how smell works, though. Its ancient evolutionary bond with the hippocampus gives the olfactory system direct access to the memory system, without having to go through the thalamus,[60] like someone with a VIP pass cheerfully skipping past the long queues and heading straight to the roped-off area of a trendy nightclub. Without the translation and relaying of signals via the thalamus, the sensory information from smell would understandably be more potent and significant, from the hippocampus's perspective.

It works both ways, too. Recent studies have shown that the hippocampus links to the anterior olfactory nucleus, a well-known but still relatively poorly understood part of the olfactory network. The hippocampus seemingly activates this region when a memory for odour is recalled.[61]

It's a very complex process, but essentially when we remember a smell, we aren't just triggering the memory of that smell. Due to the hippocampus's special links to the olfactory networks, it's more like we're literally smelling the smell again. Maybe not to the same extent as if our olfactory receptors were again being triggered by that particular odour, but it's more salient than what we experience when recalling a particular sight or sound.

This also means it's easier for the brain to reactivate the memory for when that smell was first encountered, via all the synaptic connections linked to it. And so, once again, smell acquires an advantage over other senses when it comes to forming and triggering memories.

Finally, smell influences memory so powerfully because it's also heavily intertwined with the brain's emotional processes.[62] And, as we've already seen, the memory system is heavily influenced by emotions.

Some scientists point out that smell is the sense that has the most overlapping properties with emotions, i.e. they can be positive or negative, and of varying intensity, while our other senses are more diverse and complex (except for taste, which is the poor relation of the primary senses, and largely dependent on smell[63]).

Certain smells can reliably induce particular emotional states in humans,[64] regardless of the situation they're currently in. Conversely, your current emotional state can alter or skew your perception of smells. Experimental evidence reveals that if you're told a smell will be disgusting, you're likely to find it disgusting. Similarly, if you're told a smell will be pleasant, you'll find it appealing. The fact that it's the same neutral smell in both instances is something your brain often fails to spot.[65]

Our olfactory system also provides a way of *communicating*

emotions between individuals.[66] Numerous studies have shown that if you inhale the sweat secreted by people in an emotional state like fear, you too will experience a degree of fear. And, as seen earlier, humans shed psycho-emotional tears when crying, which, if inhaled, influence the emotional states of those around us, to a degree.

All of this suggests a strong connection between the olfactory and emotional processes in our brain, as well as that between olfaction and memory. And there is indeed such a thing. Activity in the olfactory system has been shown to directly affect the amygdala,[67] the go-to hub of emotional responses, and there is a great deal of neuroanatomical overlap between the various parts of the limbic system responsible for smell and emotion processing.[68]

In fact, a part of the olfactory system called the piriform cortex, believed to be responsible for the actual processing of odour information, *includes* the amygdala, as well as related hippocampal regions. The amygdala is *part of* the olfactory system, not just linked to it,* which can't be said for the brain regions responsible for other primary senses.

Again, this makes evolutionary sense. If navigation is the precursor to memory, then emotion is the precursor to cognition and thinking, as we've seen. So, once primitive creatures obtained a sense of smell, they would need to know what to do with this new information: i.e. that if they smelled something bad and dangerous, they should get away from it. Basically, they should experience fear. If smell was the first sense, then many believe fear to be the first emotion.[69] And fear is something the amygdala is very well known for handling. So, there we have another connection that stems from our deep evolutionary past, this time

---

* This isn't to say that the amygdala is 'responsible' for our sense of smell. It's best not to think of hard functional boundaries when it comes to the brain. It's more like a Venn diagram, but one made up of thousands of overlapping circles.

between smell and emotion. And this connection still affects us today.

A wealth of data shows that memories triggered by smell invariably include a greater amount of emotional content than other types of episodic memories.[70] People who develop anosmia, the inability to smell, have often reported memory problems, and even sometimes a stunted emotional range, compared to before they lost their sense of smell.[71] Those with conditions like schizophrenia and depression also often display reduced olfactory functioning[72] (i.e. loss of smell), highlighting the close links between emotional and olfactory processing in the brain.

So deep is this connection between smell, emotion, and memory that it's also had a significant *literary* impact. *In Search of Lost Time* is a seven-volume novel by the celebrated French author Marcel Proust. Its main theme is involuntary memory, where the narrator recounts moments of his life that he's unexpectedly reminded of by external encounters and sensations beyond his control.

The most widely referenced example, sometimes referred to as the 'Proustian moment',[73] occurs very early on in the book, where the narrator dips a madeleine (a traditional French cake-biscuit thing) into his tea to soften it. Upon taking a sip of his madeleine-infused tea, he's hit by a flood of forgotten memories of visiting his aunt as a small child, and sharing her morning madeleine and tea.*

As mentioned earlier, when it comes to the perception of flavours, it's our sense of smell which is dominant, not taste. So, this pivotal moment in twentieth-century literary history is a direct result of the way that smell is fundamentally linked to the memory and emotion systems in our brain.

* For fellow Pixar fans, the pivotal scene in *Ratatouille* is clearly a visual portrayal of a Proustian moment.

I'm not suggesting that because I've actually *explained* that fundamental interplay between smell, emotion, and memory, that this book will be even more successful and influential than Proust's.

But, you know. Can't rule it out. If that did happen, though, it would undoubtedly be music to my ears.

Speaking of which . . .

Playing your song

Looking at my feet makes my grief slightly worse.

I have weird feet. They've no visible arch, they're basically flat slabs with toes on the end. In university, my housemates banned me from walking around barefoot, as my exposed feet genuinely disturbed them.

This surprised me. Before then, I'd assumed my feet were normal. Why wouldn't I? They're the only ones I've ever known. And my father's feet were similar, as are those of many of my relatives. Flat feet are, apparently, a quirk of the Burnett genome.

Unfortunately, this means that when I now see my feet, I'm reminded of my father, and therefore his passing. It's a surreal aspect of grief that nobody warned me about. Admittedly, it may never have happened to anyone else.

Stuff like this is likely inevitable, though. We know now that when we see a certain thing, our brains reliably trigger emotional memories about a person associated with it. And given how genetics works, we all have physical traits in common with our parents. This, I've discovered, can be tricky when you're mourning one, and are emotionally sensitive to any reminder of them.*

Harsh as it may seem, it's made me appreciate the *differences* between my father and me. Each thing we don't have in common

---

* It's certainly made shaving a strange experience.

means one less thing that can unexpectedly poke me in the emotional sore that is my grief.

My father loved sport, while for most of my life I've been largely indifferent to it. Conversely, Dad had no interest in science, while I'm, blatantly, the opposite. Dad was a car enthusiast, but as long as they get me where I need to go, I've little interest in them.

And Dad loved music. He had a keen ear for it, embraced it enthusiastically, even once worked in the music industry. Me? I like music well enough, but it doesn't affect me as deeply, or as powerfully, as it did Dad. Or, if we're honest, almost everyone else.

People often talk about their favourite albums, the best live gigs they've attended, about creating the ideal mixtapes or playlists for the right person/occasion/activity. And when they do, I just smile, nod, and hope nobody asks me anything, because I've honestly got no such experiences of my own.

I know I'm the oddity here. Music, and the love of it, saturates our culture. The Burnett family is full of enthusiastic singers; we're known as the 'Von Craps' back home. Nonetheless, I've always lacked the *emotional* connection to music that most people experience.

And on paper, that strong emotional connection does seem strange. After all, music is just a sequence of noises. Carefully arranged and artfully presented noises, sure, but still just sounds, mere vibrations in the air hitting our ears. What's there to get emotional about?

I'm not alone in pondering this. There's much research into the emotional impact of music in the brain, with many scientists also wondering why it causes us to feel emotions.[74] But perhaps the better question is *how* does music cause emotions? What parts of the brain are being stimulated by music, to produce such an intense emotional response? (In people other than myself, of course.)

According to the evidence, music affects the brain on several different neurological levels, from the most fundamental reflexive processes to the most sophisticated and complex cognitive mechanisms, often at the same time. Hence music can provide such an immersive, affective experience.

At the most basic level, music affects us via the brainstem, the region at the bottom of the brain that handles most of our immediate, unthinking reflexes, like blinking, or involuntary spasms of laughter. Reflexive brainstem processes are often triggered, almost immediately, in response to our brain sensing something that may be significant: something potentially beneficial, or harmful. Accordingly, our auditory cortex, which processes our hearing, has a direct link to the brainstem.[75] This means that the moment we hear something that *may* be important, our body immediately reacts – tensing up, flinching, going on alert, diverting our attention to whatever it is[76] – all via the actions of the brainstem.

When certain sounds stimulate us, they cause arousal, an integral element of affect, thought by many to be the raw material of emotion in the brain (as discussed in Chapter 2). Influencing us at this fundamental level usually involves sounds that are sudden, loud, dissonant, or feature fast temporal patterns.[77]

Sudden sounds make sense: many of our instinctive attention mechanisms are drawn to any unexpected sensory change. If we're alone in a quiet house and hear a noise from upstairs, we're immediately on edge (i.e. aroused) and focussed on it.

Loud noises, like bright lights and pungent smells, dominate our sensory processes, crowding out other things, so our brains automatically react to and focus on them. If you're holding a microphone and stand too close to a speaker, causing an almightily loud shriek of feedback, you don't ignore it; you move away very quickly, to make it cease.

Likewise, the more rapid the music, the more arousing we

humans tend to find it, often in a positive way.[78] Many listen to such music while running or exercising at the gym. Your preferred fast-paced music does indeed excite and motivate you, on a fundamental level. Evidence shows it can genuinely enhance performance at tasks.[79] So, you can tell your boss that next time they whine about the radio being on in the office.

In turn, dissonant sounds lack harmony; when heard together, they 'clash', rather than complement or blend together. We reliably find them jarring and unpleasant, hence the often chaotic, clashing sounds of free jazz are, shall we say, an acquired taste.

More typical examples of dissonant noise would be the sounds of building and construction work, of a three-year-old child hammering at a drum kit, or, the most classic example, fingernails down a blackboard. These are 'jagged', dissonant sounds, so they have a powerful effect on our brainstem processes, which results in the bone-deep shuddering reaction they reliably provoke.

However, evidence shows that not all dissonant sounds induce this unpleasant subconscious reaction; they must fall within a specific range. Basically, we dislike sounds that clash, but only if they *don't* differ too much.[80] Think about it: a trumpet and a drum produce very different sounds, but we can listen to them being played together quite happily. They sound too different; we recognise them as separate things working together. But when clashing sounds are closer together in the audio spectrum, that's what sets our teeth on edge.

Most music is easily complex enough to include sudden changes in sound, increased volume, rapid tempo, and harmony or discord. All these induce arousal, via the brainstem, so music has a direct way of stimulating an emotional reaction in the brain. And studies have shown that loud or dissonant music increases the heartrate of a sufficiently developed foetus, while soft, harmonious music lowers it.[81] Basically, music can affect us via the brainstem mechanism before we're even born.

This is a very fundamental, relatively simple mechanism, though, and there are many more sophisticated routes via which music can induce emotions in the brain. One is emotional contagion,[82] where we react emotionally to music because the music itself has emotional qualities which we can recognise, and experience ourselves in turn.

Slower music is often perceived as sad, while faster, rapid music is happier, more excitable, as most pop music demonstrates. Louder music with sudden and stark changes, which underlies most heavy metal, seems aggressive and angry. And deeper sounds feel ominous, instilling fear and dread, as the classic *Jaws* theme demonstrates very effectively.*

This ability of humans to detect an emotion, and subsequently experience it, is believed to be the work of mirror neurons, arguably one of the most important neuroscientific discoveries of recent decades.

In a landmark study on macaque monkeys in the 1990s,[83] neuroscientists were studying neurons in the motor cortex, the part of the brain responsible for the conscious control of movement. This study discovered that certain neurons in this area were activated by the monkey observing the movements of another monkey, while the test subject did nothing. It was a case of monkey-see, monkey-*not*-do.

These neurons become active when *observing* functions associated with that brain region, rather than *performing* them. They mirror the activity of others. Hence, mirror neurons.[84] Since then, activity suggestive of mirror neurons has been reported throughout the

---

* When you add key changes, verse and chorus differences, lyrics, and more, a single song or tune can have multiple emotional qualities, even ones you'd think would be contradictory, in the same way that many a slow ballad can be quite uplifting.

human brain,[*] particularly in the premotor cortex, the supplementary motor area, the primary somatosensory cortex, and the inferior parietal cortex.[85] These regions are integral for movement, language, sensation, and, in many cases, emotional reactions. Particularly the inferior parietal cortex, which allows us to recognise the emotional elements of human posture and facial expressions.[86]

Mirror neurons are believed to underlie the process of empathy,[87] which makes sense: neurons that mimic the activity you observe would be extremely useful for recognising people's emotions and prompting the same activity in our own brains. And given how we readily hear emotions in someone's voice, via tone, delivery, etc. without ever seeing them, empathy clearly occurs via the auditory system too.[88]

And this process can be triggered by music: mirror neurons, in the cortical sensory regions of our brain, detect the emotional component of the music and cause us to experience it ourselves. This is emotional contagion.

But there is also another, more cognitively complex mechanism via which music induces emotions in the brain: musical expectancy.[89]

Music has structure, patterns, themes, and an underlying grammar in the form of music theory. Verses and choruses, build-ups and crescendos, chord structures, modes, time signatures, and many more complex things that a tin-eared amateur like me doesn't even recognise. In this sense, music is like language.[90]

We readily manipulate language to induce potent emotional reactions. Exquisitely timed jokes or wordplay can cause happy laughter; a well-structured poem can instil profound sadness; a clever narrative can cause us excitement, dread, or apprehension;

* Specific mirror neurons haven't been identified in humans yet. We don't really have the technology to observe activity in a specific neuron in a living human brain. But there's still much compelling evidence for their presence regardless.

and so on. Similarly, deft manipulation of musical structure and convention also inspires emotional responses. This is musical expectation, where you have a certain level of understanding, or expectation, of music. When a piece of music meets, or preferably *exceeds*, this level, you experience a positive emotional reaction.[91]

Conversely, if the music falls far below the standard you expect, you get a negative emotional reaction. Musical connoisseurs are sometimes disdainful and dismissive of 'mainstream' music, often because it's mass-produced and commercial, hence targeted at the lowest common denominator. To their ear this music might seem unsophisticated, so doesn't stimulate their musical expectation.

This phenomenon is supported by evidence showing increased activity in Broca's area – the higher neocortical brain region responsible for much processing and understanding of language – when we listen to, and appreciate, music.[92] This finding suggests that our cognitive brain systems are as involved as the lower instinctive ones when it comes to the emotional impact of music.

However, musical expectation varies from person to person. It's like wine: an experienced sommelier with a very refined palate can apparently recognise a sauvignon, a pinot, a chardonnay, and which year they were bottled. They can appreciate the variances of different grapes, and the subtle qualities like peachy aromas, oaky finishes, hints of pear and asparagus, all that. Then there are people like me, who can tell between red, white, and rosé, and that's about it. I like wine well enough, but the subtle complexities of it are totally lost on me.*

The same applies to music: if you've developed the palate to appreciate its more refined aspects and properties, you presumably get a lot more from it, on a cognitive and, subsequently, emotional

---

* Although see my first book, *The Idiot Brain*, for the somewhat contentious relationship between wine tasting and neuroscience.

level. Musical expectancy develops and grows in response to the extent to which your brain is exposed to music. And it's never too early to start, as some studies report that playing music to babies in the womb leads to advanced recognition and appreciation of more complex music in early childhood.[93] Music appreciation, and your emotional reactions to it, can end up in something of a positive feedback loop, as the more you listen to and enjoy it, the more able you are to enjoy it in future.

That's presumably why free jazz or heavy metal, with all their discordant elements, are still enjoyed by those who can cognitively appreciate the skill and complexity they display, and so overrule the more primitive aversions they may trigger via their brainstem.*

Scanning studies have also indicated the role of memory in our enjoyment of music, showing that the more *familiar* music is, the more our brain responds to it, in a positive emotional way.[94] Familiar music causes greater activity in the limbic, parahippocampal, and the cingulate cortex regions of our brain, all more established, lower regions of the brain, responsible for, or involved in, emotion and memory processing.[95, 96]

The point is, rather than growing bored with it, as usually happens, we like music *more* if it's familiar. That's why we can listen to songs on repeat, why we always want to hear 'the classics', why we often prefer one particular genre of music over the others. Because we like that which is recognisable and *remembered*.

Interestingly, this happens both on a conscious and an unconscious level. For instance, have you ever experienced an inexplicable emotional reaction to a song? It's not the sort of music you usually like, it's not especially complex or impressive, you may even find it annoying, but you can't help but find

* Although, even if something seems objectively bad, you can still enjoy it on an emotional level. Remember all that stuff about spicy food and BDSM?

yourself liking and enjoying it anyway? Personally, I'm quite fond of Vengaboys, the late-1990s Europop band who informed us, repeatedly, that the Vengabus is coming. I'm fully aware that their music is just repetitive cheesy fluff, and yet it still stirs positive emotions in me anyway. What's that about?

The most likely explanation is what's known as evaluative conditioning,[97] which is when our otherwise neutral feelings about something are changed because of an experience where the thing in question is associated with something we actively like, or dislike.

For example, you may be utterly ambivalent about Westerns, but go on a date with a man who's a huge fan of them and wants to share his enthusiasm with you. You end up falling in love and getting married. And now you quite like Westerns, because they're indelibly linked in your brain – in your memory – to a source of immense happiness.

This can easily happen with music: if we have an emotional experience with a song playing in the background, our brain automatically links them together in our memory. And so, later, when we hear that song again, it triggers an emotional response via this memory connection.

This has a disproportionate impact on our emotional relationship with music because incidental music pops up everywhere in the modern world: the car radio, street buskers, background music in shops, bars, hotels, and so on. As a result, experiencing emotions while hearing music is a very common occurrence, meaning our brains regularly connect the two, because of evaluative conditioning.

It's very much an unconscious process, occurring via limbic and lower regions like the amygdala[98] and cerebellum.[99] In fact, some evidence suggests that awareness of the music playing as you're experiencing emotions actually *hinders* the association process.[100] And this association, between music heard and the emotion experienced, is surprisingly stubborn, when compared to similar unconscious associations between other types of experiences and stimuli.[101] Essentially, when your brain connects music to a certain emotion, it's very reluctant to undo this connection.

So, if you've experienced a positive emotional response to a song you wouldn't expect to, you may have simply overheard it while in a good mood. In my case, the music of Vengaboys was a constant presence during the period of my teens when I suddenly went from shy and timid to far more confident and outgoing, and started enjoying life more as a result. My memory has, understandably, pinned Vengaboys to this time, hence I'll always have a soft spot for them.

Overall, if you've ever wondered why we have musical 'guilty pleasures', where we like songs we feel we shouldn't, it's probably due to evaluative conditioning.

Which brings us to the more obvious, conscious role of memory in the emotional experience of music, which occurs via our old friend, episodic memory. The key difference between this and evaluative conditioning is that the latter occurs when the music is in the background, incidental to whatever it is we're doing.

But when we're consciously listening to music, be it in our room with headphones, or attending a festival we've spent months looking forward to, we're embracing all the emotional reactions it creates in us, and the episodic memory processes are engaged. Because, as repeatedly demonstrated, emotional experiences boost our memory system, meaning we're far more likely to remember them than non-emotional ones.

So, music that stimulates us emotionally (whether we like or hate it) is far more likely to be consciously remembered. And, because memory and emotion have a two-way setup, remembering music causes us to feel the emotions associated with it all over again. It's sort of a messy feedback loop.

This perhaps explains why we're more emotionally responsive to music we're familiar with than that which we're not; familiar music has this elaborate emotional boost triggered via the memory system. Novel music, something not found in our memory, does not.

It also explains evidence which shows that, like smell, music tends to trigger more strongly emotional memories.[102] This is believed to be because listening to music induces emotions via multiple neurological mechanisms, as we've seen, so more emotional elements are included in the eventual memory of it.[103]

This potent connection between music and emotion and memory can have some unusual effects. For example, earworms, the common phenomenon where you can't seem to stop replaying a song in your head, even if you don't like it and it's actively annoying you.

There's a surprising amount of research on earworms, but still no definitive answer on why they happen. Some point to their similarity to ruminative thoughts, i.e. not being able to stop thinking or worrying about something that's bothering you. This implies that people who are stressed or anxious might be more prone to earworms – a notion supported by some of the research[104] – which in turn suggests the involvement of emotions in the earworm process.

Others describe the particular traits of music that readily lead to earworms, highlighting that they often rhyme, tend to be simpler, repetitive, harmonious, or have a sort of 'loop' structure that suggests no obvious endpoint, so the brain can, and will, keep playing them over and over.[105]

Often earworms don't even need to be heard to be triggered; a simple cue from something vaguely related in memory can set them off. But whatever the underlying mechanics, it seems that earworms are pieces of music that stimulate the memory and emotion system in just the right way to end up being constantly recalled, to an often infuriating degree.

A more profound example of how music, memory, and

emotions interact is the fact that people mostly prefer the music they were exposed to in their youth, especially their teens.[106] This is the phenomenon of the 'reminiscence bump', where no matter how old you get, the memories from your adolescence and early twenties tend to remain clearer than others.[107]

Many neurological factors contribute to this, like the fading affect bias, gradually removing all the bad emotions in our older memories, leaving only the good emotions behind.* Also, our higher brain regions, those responsible for much self-control and cognition, don't finish maturing until our mid-twenties, while the simpler emotional regions are ready to go much earlier.[108] This means the parts of our brain that help keep emotions in check are still developing during our adolescence, which is why our emotions are so much more intense during our teens.

Accordingly, the memories from our teen years will have a greater emotional component, and therefore be easier to recall than those formed later in our lives, when our cognitive processes have our emotions on a tighter leash. And because we're 'more emotional' during our teens, music that emotionally stimulates us will have a bigger impact on us than at any other time.

Also, during our teens we actively, instinctively seek out peer approval, acceptance, and novelty, so will explore more new sensations, like new music, to express ourselves and gain approval.†
We also listen to more emotional music to help us comprehend the confusing emotions we're experiencing, and to help us feel understood, accepted.

Overall, there are a great many ways that music can, and does,

---

* Hence many people refer to 'the good old days', even if they were terrible.

† The maturing of our emotions and reward systems means that, during our teens, things we liked during our childhood lose their potency. This means that our teens are perhaps the one time in our life when we *don't* prefer familiar music by default.

have a much stronger emotional impact on us when we're teen-agers. And thanks to how memory works, this often determines our preferences, for much of the rest of our lives.

It's weird to think that music can be such a big influence on shaping who we are, but there it is. And not just on an individual basis, but an evolutionary one too!

Studies have shown that pleasurable responses to music show very similar activity in the brain's reward systems to that displayed when we enjoy delicious foods, sex, and recreational drugs.[109] This pleasurable reward response is usually reserved for things that are biologically relevant, that were important for our own and our species' survival – or, in the case of drugs, that hijack these reward systems from the outside. This means that our emotional reaction to music has deep evolutionary roots; it must have been (and maybe still is) important for the ongoing survival of humanity.

This is widely believed to be because, as with certain colours and smells, certain sounds are associated with specific things in nature that our brains evolved to recognise and respond to on an immediate, emotional level.[110]

Maybe we find dissonance so unpleasant because the screams and hunting cries of predatory animals are often very dissonant noises, so our brains evolved to be wary of such sounds, and what-ever causes them.

Maybe we find slow rhythms sad and faster ones happy because slowness of movement and speech is a sign that someone is in a low mood, or maybe injured, while rapidity suggests excitement and energy. Perhaps this is linked to our heartbeats? On some subconscious level, we recognise that rapid heartbeats are a sign of excitement, and slow ones of calm and relaxation. A lot of pop music falls within the 100–120 beats per minute range, just above an average heartrate, which would mean it's perceived as 'energetic'.

Maybe we find rich, complex music more rewarding and

stimulating than simple, separate sounds because an environment with many overlapping sounds suggests life, resources, abundance. Meanwhile, quieter, softer music is often relaxing, perhaps because we associate it with an absence of dangers while we're still aware of what's happening around us?

It's sudden, unexpected silences that can actually be unnerving. Maybe because it triggers some ancient reflex responding to when everything in our local environment goes quiet to avoid the notice of a nearby predator. This would explain why some people find silence so unsettling. This might also be why some people need background noise (usually music) when working so they can concentrate. Absolute silence can be, ironically, disquieting.

There are even theories that suggest we perceive emotional qualities in music because instruments share audible properties with voices. The super-expressive voice theory[111] argues that the mechanism via which we recognise the emotional quality of speech is engaged by the brain when we listen to music. It's an interesting theory, although more recent research suggests the brain does actually process music and voice separately.[112]

But logically, that would only apply when our brain can readily *differentiate* between music and voice. This need not always be the case. Instruments most often deemed 'expressive' – cello, violin, slide guitar, and much woodwind or brass – are those capable of glissando or portamento, i.e. sliding between notes. This mimics the sound of singing more closely than discrete notes, and has a noticeable impact on the emotional resonance of the music.*

So, even if our brain does have distinct separate systems for processing music and voice, the rich and flexible nature of music, and the objects that create it, means that any dividing line between the

---

* Also see 'sad trombone'.

two will inevitably be rather blurred.

Another thing: why do we like harmony so much? Well, we humans are tribal creatures first and foremost. We value social-isation and interaction so much that anything that can emphasise group unity and cohesion is normally perceived very positively by our brains.[113] And what could better emphasise how united a group is than every member making the same sound at the same time? The more complex the better, as it shows how in sync and capable we are. Many contend that this is why humans respond so strongly, so emotionally, to music in the first place: among other things, it's an excellent method of uniting a group.[114]

This may even explain our urge to dance, and our enjoyment of doing so. Remember, music stimulates the motor cortex too, suggesting it causes a compulsion to move. From an evolutionary perspective, even better than a unified tribe is using that unity to actually do something. Hence, a compulsion to act – to move, but in a coordinated, harmonious manner – would be something the evolving brain would like very much.[115] And that describes dancing rather well, I'd argue.

Undoubtedly, humanity's embracing of music and dancing has developed in vastly complex ways throughout our history.[116] But it likely began with our primitive brains recognising that coordinated vocalisations and movements were things to be encouraged when trying to survive the constant dangers nature threw at us.

This exploration of the origins of our underlying emotional connection to music served to bring me back to wondering about my own. Despite all I'd discovered, I still didn't really know why I'm not as affected by music as most people.

In a way, it makes even less sense now. Musical expectation develops in accordance with how much music you're exposed to,[117] and I was exposed to it constantly growing up. Both my parents were music lovers, and we lived in a pub, where music was playing all the time, either live or via the jukebox, radio, etc. There was always somebody playing music somewhere . . .

And that's when it occurred to me: I grew up in a busy pub, but we didn't move in until I was about two. I was quite a shy, timid child from day one, so going from a small, quiet terraced house where just three people lived, to a huge draughty building with strangers wandering around must have been quite an unpleasant shock for two-year-old me.

And this environment had music playing *all the time*. Via evaluative conditioning, your brain automatically associates the music you hear with the emotions you experience. So, at this very formative time, maybe my developing brain readily associated music with big scary changes, with unfamiliar adults drunkenly wandering into my bedroom,* with nights disturbed by loud discos in the nearby events room right across from our living area.

I don't feel that way about music now, but this may have prevented or dampened all the positive emotional associations with music I would have otherwise learned at a key stage of development, that would have given me the same enthusiasm for it that everyone else has.

Ironically, if I'm right about this, my not enjoying music as much as my dad is actually his fault. It was his decision to be a landlord, after all.

But, given the circumstances, I'll forgive him for that.

I'm still a bit annoyed about the feet thing, though.

However, thinking back to all the disrupted nights and scary experiences of my earlier childhood triggered another line of emotional investigation, and getting to grips with it was a nightmare.

---

* A surprisingly common occurrence.

The nightmare scenario

In the wake of my father's death, I started having weird, unpleasant dreams.

That's as much as I'll say about them, because, let's be honest, the phrase 'Let me tell you about my dream' is one virtually guaranteed to cause the listener's eyes to glaze over, as they're consumed by a wave of extreme disinterest.

Still, I will say that having more upsetting and vivid dreams than usual was rather annoying, given all the other emotional issues I was trying to get to grips with. In some ways it is unsurprising, though: the sudden loss of a loved one undeniably makes your daily life a fog of emotional turmoil and confusion, so why *wouldn't* your dreams follow suit? It's the same brain responsible for everything, after all.

I'm not alone, either. The COVID-19 pandemic has led to people worldwide being more scared, confused, anxious, angry, and stressed than usual. The overall mental health impacts will take years to uncover and unpack, but interestingly, at the time of writing, I'm seeing a lot of people online opening up about the weird, unsettling dreams they're regularly having.

The obvious conclusion here is that the emotions we experience while awake significantly influence the dreams we have when asleep. Of course, not all emotions are good. Some are very negative, so would presumably lead to very negative dreams: nightmares.

We've all experienced nightmares. They're deeply scary and unpleasant, but also remarkably common. Data suggests that between 2 per cent and 6 per cent of us experience them on a weekly basis.[118]

Why, though? While negative emotions are useful, and often more stimulating than positive ones, our brains also work hard to suppress or limit their impact and influence, via things like the fading affect bias. So, if emotions like fear are our brain's way of knowing what we should avoid, why does that same brain *induce* fear in our dreams? Is there a purpose for this baffling function, or is it a sign of something going wrong?

Luckily, while hearing about someone's *specific* dream is tedious, dreams and dreaming as a process are the subject of much interest, including for many scientists. So, there's a lot of research out there about how dreams and nightmares work. While the specific nature and processes at work are still debated, it's widely agreed that they're an important, possibly even vital, aspect of how our brains deal with memories and emotions.

Sleep has four distinct stages: non-REM stages 1, 2, and 3, followed by REM (rapid eye movement) sleep.[119] Dreaming occurs during REM sleep (with some rare exceptions[120]). The longer we're in REM sleep, the more we dream, and the REM stage seems to last longer each time we go through the cycle in a typical night. This means REM sleep lasts longest at the end of the sleep cycle, hence we're often woken up from a dream by our morning alarm.

The bigger question of why we dream is tied into how our brains deal with memory and emotion. Dreams have an important role in memory consolidation;[121] after all, what better time to bolster existing memories than when we're unconscious, the one time when no new memories are being formed? Consolidating memories while we're awake is like repairing a road while cars are still driving on it: it *can* be done, but it's much harder.

While our dreaming brain is integrating new memories, linking them to older, established ones, it 'activates' them, to an extent, meaning we re-experience them. This happens regularly when we remember things while awake, but our real-time consciousness takes up most of our brain activity. When we're asleep, though, our consciousness and senses are largely 'shut down'. This means the memories being triggered while dreaming, and the experiences they entail, have the brain all to themselves. That's why dreams seem so 'real' while we're experiencing them: they're activating memories, but without our consciousness overshadowing them, these memories are a lot more 'immersive'.

However, we've seen that each memory* is an amalgamation of elements. A memory is a combination of specific sensory, emotional, and cognitive experiences, stored in the brain as collections of synaptic connections. This allows certain elements to be used in multiple memories, saving space and resources in the brain.

It also means that not every element of a memory need be activated when we dream. Discrete elements of memories are triggered individually, and linked to relevant elements of other separate memories,[122] strengthening the integration, utility, and usefulness of the materials stored in our memory.

This can explain why what happens in our dreams is often so bizarre: they're distinct aspects of existing memories, being combined and triggered in atypical and unusual patterns. The people you dream about are often amalgamations of other people you've encountered, the places mashups of where you've been while awake, and so on.

And there doesn't need to be any real-world logic as to which bits of memories are linked together. For instance, if you've a memory of singing and a memory of being underwater, the dreaming brain can activate both, giving us the dreaming experience of singing underwater, even though human physiology and the laws of physics don't allow that.

But even when our dreams include such outlandish scenarios, our dreaming self rarely thinks anything unusual is happening. This makes sense: if our dreams are entirely constructed from elements of memories, then our dreaming brain technically *isn't experiencing anything new*.

---

* Specifically, episodic memories, which provide the main fodder for dreams.

The pivotal role of sleep and dreaming in memory processing and consolidation has been demonstrated in several interesting studies. One, invoking the power of smell, exposed subjects to the smell of roses while they performed specific learning tasks. After this, some subjects had the rose smell pumped into their rooms as they slept in the lab overnight, while others did not. The study showed that those who were exposed to the smell in their sleep performed a lot better on assessments of learning the next day.[123]

Does this mean that if you have a scented candle in your room while you study, then have the same lit candle in your room while you sleep, you'll retain the information better? This experiment suggests so. I certainly wouldn't recommend sleeping with an open flame in your room, though.

Similarly, the advice to 'sleep on it' when faced with a difficult problem or decision seems to be a scientifically valid approach. Studies have demonstrated that people woken up during REM sleep are significantly better at complex problem solving* than those woken up during non-REM sleep.[124] This implies that, when we're dreaming during REM sleep, our brain is in a more 'flexible' state, where uncommon connections between memories and processes occur more easily.

That's arguably what you'd expect if dreaming involved the memory-consolidation processes described. The brain would *have* to modify itself to allow random, atypical connections to occur more readily. So, yes, if there's a problem or issue you're struggling with, sleeping on it could be a big help. All recent memories concerning the problem will be better integrated with your existing neurological setup, and there's a greater chance of your brain making the link between problem and solution than when we're awake and relying on our established (i.e. more rigid) neural pathways.

* This particular study used anagrams, if you're interested.

This is related to the continuity hypothesis,[125] a relatively straightforward theory which argues that the dreams we have at night are largely shaped and determined by the experiences we accumulate during that particular day. Our most recent memories are obviously the most in need of processing and consolidating, so it wouldn't be surprising if they take priority during dreams, when such things occur.

However, this logical theory doesn't quite explain why our dreams are often wildly confusing and unpredictable. But a further look at what the dreaming brain is up to can help explain these issues.

For instance, the hippocampus is, if anything, even more active during dreams,[126] further emphasising that dreaming has important roles in memory processing. But hippocampal activity in dreams is different to that seen in a waking brain.[127] It's widely agreed that this unusual hippocampal activity is the reason dreams are so surreal and weird. The hub of our memory system is behaving abnormally, so the memories it ends up triggering are similarly weird. It's also argued that this is why dreams are so hard to remember: the hippocampus, essential for memory formation and recent memory recall, is behaving differently when we're awake, and this distorts and compromises our ability to retrieve memories formed when in its 'other' state.

So, the link between dreaming and memory seems well established; but what of emotion?

Although the theories about the actual mechanisms are many and varied, it's generally agreed that dreams enable us to process emotions and emotional experiences. This notion is far from a modern invention: Sigmund Freud wrote an influential book about dreams and their meanings 120 years ago.[128] Freud's contention was that dreams are the brain's way of sustaining sleep by containing and processing the anxiety caused by our sexual urges, which would otherwise wake us up. Nightmares, in turn, are when our sexual urges take on a more masochistic bent, causing dreams that are very unpleasant and upsetting.

While most modern psychoanalysis has moved on from pinning everything on disturbing sexual urges, there is still consensus that dreams are a key aspect of emotional development, and that nightmares are a particular expression of this. Or, they're where the process goes awry.[129]

Neurological evidence backs this up. As well as the hippocampus, the amygdala is also more active during REM sleep than when we're awake,[130] which suggests that, whatever's happening in dreams, emotions are a very important part of it.

This all makes more sense when you remember that the emotional component of a memory is a separate element, in its own right. Amygdala disruption or damage can cause memories to lack any emotional aspect, while the other details of the event are preserved in memory. Clearly, the emotional experience can be isolated and separated.* And this is, in essence, what's happening in dreams.[131]

We all know that our emotional experiences can stay with us, and continue to affect us, for a very long time indeed. This suggests that the memory of the event is still triggering the associated emotion, and because of the complex, reciprocal nature of how memory and emotion work in the brain, this can create a feedback loop. The emotion activates the memory strongly associated with it, which further triggers the emotion, which activates the memory again, and on it goes.

As we've seen, such things as listening to sad or angry music can provide a way to process the stubborn emotion we're grappling with, triggering it without necessarily activating the potent

---

* It may seem like this contradicts the whole 'emotions and cognition are impossible to separate' conclusion from Chapter 2, but there's a big difference between how things are produced in our brains and how they're perceived by us and stored in our memory.

memory/memories associated with it. This allows other parts of our brain to engage with the emotion, to form links and associations of their own, meaning the potent emotion is somewhat 'detached' from the memory of the event that caused it, and its effects are spread further throughout our brain, lessening its potency and increasing the brain's ability to 'deal with it'.

This, in many ways, is exactly what's happening when we dream. Our brains effectively take the emotional elements of memories and link them up to other memories with similar emotional features, to enhance future recognition and processing of such emotions. In so doing we reduce the potentially disruptive influence of the original emotional memory.

Many feel this is one of the more important functions of dreaming, which presumably explains my own altered dreaming tendencies, and those of others struggling with life amidst a pandemic. Our emotional state during the day and before we go to bed has a potent impact on what we end up dreaming.

So, if you're stressed, due to work or your relationship or whatever, your sleeping brain will take these emotional elements from your newly formed memories and link them up with memories that have similar qualities. Even if you consciously suppress the stress during the day, the subconscious brain is still very much aware of it, so tries to tackle it in our dreams. And so the negative emotions it contains spread out through our dreams.

And here's where it gets particularly interesting. While many (including Freud and co.) believe that nightmares are a common and inevitable aspect of dreaming, more recent theories from the world of evolutionary psychology and beyond argue that nightmares are actually *the whole point* of dreaming. Or at least they were originally.

For instance, the threat simulation theory[132] suggests that dreams first evolved as a means for our brains to simulate threats and hazards while we slept, so we'd be more able to deal with such things if they occurred in the real world, as we'd already had 'practice', of sorts. More complex and emotionally diverse dreaming developed from there as our brains became more advanced. This would mean that our brains first started dreaming to learn to engage with, and figure out how to avoid, things which caused fear. So, we're meant to have nightmares; it helps us survive.

Supporting this is the fact that we have way more nightmares when we're younger.[133] Our youthful, inexperienced brains haven't worked out how to recognise and deal with dangers yet, so need to run more simulations, i.e. have more nightmares.

It's an interesting theory, but many (myself included) don't think it entirely holds up. The obvious counterargument is that it suggests that the more nightmares we have, the better our mental health should be, because our brain is logically spending more time processing, and dealing with, things that cause stress and anxiety. But that's not the case at all.

Many mental health problems, particularly those of an emotional nature, are associated with increased occurrence of nightmares.[134] Indeed, psychologists recognise two different classes of nightmares: idiopathic (the kind most people have every now and again) and post-trauma nightmares, which occur very often and in great intensity when someone has experienced a profound emotional trauma.[135]

You'd obviously expect someone suffering from PTSD or similar to have copious negative emotions in need of processing. If nightmares are a means of doing that, you'd expect them to be helpful with this. Instead, increased and recurring nightmares are often a sign of mental and emotional health decline, whereas reduced nightmares are a sign of better long-term recovery.[136]

Could it be that nightmares are both unhelpful *and* necessary? It might sound contradictory, but we've seen repeatedly that the human brain is more than capable of that sort of versatility, particularly where emotions are concerned. Whether nightmares are helpful or disruptive could be more down to context and circumstance, like many other things emotional.

Indeed, numerous modern ideas about how and why nightmares happen seem to adopt this 'both a blessing and a curse' perspective. One good example of this is the affect network dysfunction (AND) model of nightmare production, which suggests that they occur because fearful memories are particularly hard to extinguish.

Our brain is, understandably, loath to forget anything that scares us. The whole point of experiencing fear is to teach us that something's dangerous, so remembering what scares us has long been essential for survival. As a result, fearful memories are particularly stubborn, hard to forget. And even if we do manage to forget them, they can easily be 'reactivated' later.[137]

However, having memories that cause a powerful fear response just lurking in your brain isn't good either, as anyone with PTSD will know only too well. The AND model argues that this is where bad dreams and nightmares come in.

Rather than suppressing or removing the existing fear-infused memories, they're a way of effectively 'replacing' them. Bad dreams and nightmares are our brain's way of detaching the powerful fear element from upsetting memories, and attaching it to other ones that aren't as evocative. By doing so, it lays numerous 'new' memories, which incorporate the tricky fear experience, over the top of the original one.

These new associations with the intense fear aren't as powerful and robust as the original waking experience, though. They're not as stimulating, and we've seen that dream experiences are harder to remember as it is. Therefore, it often requires multiple 'attempts' at covering over the disruptive memory with new memory combinations during the dreaming process to make it really 'stick'.

It's a bit like moving into a new house which has wallpaper with unpleasant patterns that you don't like. So, you paint over it. But the pattern is very boldly coloured, and the paint is thin, so it takes several coats to truly cover it up. The new memory combinations formed during dreams are similarly 'thin' compared to the powerful waking memory, so it takes several attempts to truly overwrite the troublesome memory.

This would explain why recurring dreams (or nightmares) are a thing, and why we can have similar dream experiences several times in a night during the separate sleep cycles; our dreaming brains are trying to tackle a particularly potent emotional memory, and it takes several REM cycles to do so effectively.

However, this process seems quite a delicate one. It's relatively easy for it to become overloaded, given how powerful negative emotions can be. This is made apparent by the fact that one of the defining features of nightmares is that they cause the individual to wake up.[138] That's very telling: we need to be asleep for our brain to process emotional memories in healthy manner. Nightmares stopping us from sleeping can't be a viable part of the process, so presumably something has gone wrong there.

Indeed, those who suffer from post-trauma nightmares regularly report chronic sleep loss* and disturbance, and both classes of nightmare are associated with increased limb movement, suggesting the brain and body aren't as 'asleep' as they should be. All this has led to many arguing that nightmares should be classed as a sleep disorder in their own right, rather than a symptom of other anxiety or emotional problems.

Taking all this into consideration, it now looks like nightmares *are* a sign of something going wrong after all. Dreams

---

* Which is partly why things like PTSD are so enduring and disruptive; sleep is when the brain works on sorting out the problem, and sleep is being lost as a result of the problem.

are certainly an important feature of our sleeping brains; they're where our memories and emotions are properly processed while the brain isn't busy with other things. Dreams involve elements of our memories being activated separately and combined with other memory elements in new and unusual combinations, hence our dreams are so often bizarre and nonsensical. But this allows the emotional experiences embedded in our memories to be spread throughout and better integrated with the rest of the brain, which is why our dreams can be so emotional in nature.

But nightmares interrupt this vital process. They're so scary, so intense, that they regularly cause the brain to abandon sleep and dreaming altogether, so the emotional build-up in our memories remains unprocessed, causing further problems.

So, nightmares do indeed seem both necessary *and* unhelpful. But perhaps this isn't as contradictory as it might at first seem: what if we distinguish between bad dreams and nightmares? Bad dreams, it seems to me, are when our sleeping brain deals with negative emotional memories effectively and successfully. Nightmares are where they try to do that, but fail, because there's just too much negative emotion in the memories being worked on, and the brain's capacity to handle it is overloaded.

Life is often very emotional, so it's unsurprising that many people will have nightmares from time to time. That children and teens, with brains that produce more potent emotions but have less practice at processing them, would have more nightmares makes sense in this context.

It provides some reassurance for me, at least. As troubling and unenjoyable as my dreams have become since Dad died, I've not been woken up by them, thus far. I think I'm still on the 'bad dream' side of the equation.

I've just lost a parent in traumatic circumstances, and because of the whole pandemic thing, I've been kept away from all my family and friends for months. That I have a glut of negative emotions to deal with is undeniable. But, thus far at least, they've not proved too much for my unconscious brain to deal with.

Unless, of course, there really *is* something amiss with the emotional wiring of my brain? But let's not go down that road again just yet. I figure I've got enough to deal with right now as it is.

I will say, though, that, while I would obviously prefer everyone to sleep soundly and peacefully every night, it was reassuring to know that I wasn't alone in experiencing bad dreams in the wake of tragic and upsetting experiences. Other people revealing they were going through similar experiences was oddly comforting.

Relatedly, my own emotional confusion during these dark times has undeniably been lessened by the act of writing it all down, and sharing it with you, unknown reader.

This led me to an obvious conclusion: experiencing emotions is only part of the process. For us humans, *sharing* them, communicating them to others, is often just as important a part of our emotional existence.

4

Emotional Communication

Grief is hard. Everyone knows that. So, the fact that it was such an emotionally challenging experience wasn't at all surprising for me.

What *was* surprising, though, was the *variety* of ways in which it caused havoc with my emotions. I'd assumed it'd be a long period of intense sadness, like on TV. It's not, though. Immediately after my father's passing, I mostly felt numbness. When the intense sadness did eventually arrive, it came in waves, including occasional periods of low-key weeping, but punctuated with bursts of anger and frustration, which seemingly popped up apropos of nothing.

Some days, I felt . . . fine. Good, even. But then I'd feel guilt and shame about that; my father had just died, and here I was being upbeat? How callous of me! As if I wasn't emotionally confused enough already.

Eventually, I started to worry that maybe I wasn't grieving 'properly'. This whole journey started with my concerns that my own emotional processes had gone awry in some way, and my apparently bizarre grief experience would be consistent with this.

The problem was, this was my first experience of something like this. So, I didn't really know how it was supposed to go, how my grief was meant to 'pan out'. I maintain that I'd have had a much better idea if I were around my family, or friends of Dad's. They were grieving too, so we could relate, talk it out, share our feelings, reassure each other, and so on. That's how it usually goes, after all: you experience a loss, and those close to you rally round, to console you, to share your pain, help you get through it.

I couldn't do that, though. I had to deal with my grief alone. Everyone was under lockdown, and given how the virus had just taken my father, I took that very seriously, staying away from friends and family, no matter what it cost me.

But . . . perhaps it cost me too much. Because when it comes to emotions, as well as everything else they do, an obviously important part of their function involves expressing them openly, in ways that others can recognise. Why else would so much of our brain and body be dedicated to, and shaped by, the ability to portray, detect, and *share* emotions?

Indeed, a surprisingly large component of our emotional experience is made up of the feelings and emotional reactions of others. Without them, our own emotional life is diminished, like watching a film with the colour removed.

My concern was that this was happening to me, that being deprived of contact with those I care about, at such an emotionally fraught time, was compromising *my own* ability to experience and process grief. And not just that, the other emotions too. Positive ones, that could help stave off the bleaker feelings. It's surely much easier to laugh about the good times when you've someone to laugh with?

So, having little else to do, I figured it would be wise to investigate whether this was a valid concern, by uncovering what the science actually says about just how important the communication, the sharing, of emotions is, for who we are, and how we work.

It turns out, the answer is 'very'.

I feel your pain: empathy, and how it works in the brain

In truth, I wasn't *completely* alone with my grief. Yes, I was stuck at home and cut off from my extended family at the most emotionally painful period of my life. Luckily, I live with my wife and two small children. And I couldn't have managed without them.

Nonetheless, it still often felt like I was dealing with my grief alone, because I chose to keep my feelings to myself. Granted, that sounds dumb, masochistic even, but I had valid reasons, beyond 'macho posturing'.

At the time, my children were still very young. Even at the best of times I'd be very reluctant to dump a vat of adult-strength grief over them. And those *weren't* the best of times. The pandemic had deprived them of school, friends, family, travel, and, most importantly, their beloved grandfather. The idea of burdening them with my grief *on top of that* filled me with horror. So, I didn't.

Then there's my wife, the most intelligent, generous, and competent human imaginable. She repeatedly said that, whatever I needed, she was there for me. However, me churning out words in my outdoor office is how we pay our bills, which means she runs our household, and is primary parent for our children. Combined with her own career, she already has the equivalent of at least three full-time jobs. With our children indefinitely stuck at home, her workload increased drastically.

So, while I knew she meant it when saying she was there for me whenever I needed, I honestly couldn't bring myself to burden her further. Her wellbeing is as important to me as mine is to her. To have her act as my sole grief-sponge, on top of everything else she's dealing with, would have caused me such guilt that I'd have inevitably felt much worse. So, what would be the point? Instead, I opted to deal with my grief alone, and worked hard to convince my family I was doing fine.

Only, I didn't fool anyone; my wife was clearly able to tell when grief was hitting me hard, and subsequently distracted the kids, took care of things, and bought me the necessary space to ride out my emotional turmoil. My son also did his bit, dispensing hugs when needed, toning down his youthful enthusiasm, and just being as considerate as you could realistically expect an eight-year-old to be. Even my daughter sensed my moods and tried to help. Although, as an extremely forthright four-year-old, her approach involved shouting 'BE HAPPY!' at me, followed by a confident thumbs up.*

Looking back, it's *good* that my attempts to disguise my grief from my family didn't work. It would undoubtedly have done me serious mental and emotional harm to deal with it entirely by myself. But even so, that I failed so pathetically is very revealing. Despite my best conscious efforts to hide my emotions, I was demonstrably still broadcasting them to the world, in ways that even a small child recognised.

What this reveals is just how deep and fundamental the communication and sharing of emotions is for us humans.

We all know the cliché about how a great deal of human communication is nonverbal, that words and language are the metaphorical tip of an interaction iceberg, the visible part of a much greater bulk of subconscious communication below the surface. Much of this subconscious communication is tied to emotions. That's why, even though our *understanding* of emotion is vague and nebulous, *communicating* our emotions is surprisingly easy. We regularly do it without even trying.†

It's not that you can't use language to convey your emotions, because obviously you can. I could say, 'I am happy/sad/angry/scared, etc.' to any random person, and they'd understand how I'm feeling.‡ However, you don't *need* language to communicate emotions, because our brains are very good at detecting and deciphering any

---

* The fact that this constantly reduced me to hysterical laughter convinced her it works. Technically she's not wrong, but it's clearly a bad precedent.

† Or, as in my case, actively trying to prevent it.

‡ They wouldn't necessarily *care*, but they'd understand.

sensory information that's suggestive of emotion. And we humans generate a lot of such information. The chemicals in our sweat and tears; the tone, volume, pitch, and speed of our voice;[1] our laughter or cries of frustration; our posture,[2] body motion,[3] gestures, or faces (via colour and configuration). We're perpetually, albeit often unwittingly, broadcasting a wide range of multisensory cues that tell others what emotions we're currently feeling.

But detecting and recognising other people's emotions is just the start. In many cases, we *share* them. Seeing that someone's feeling down often makes us sad. Someone being frightened often triggers our own sense of fear and apprehension. We laugh way more when around others who are laughing.[4] All this, and more, demonstrates *empathy*, the ability to understand and share the feelings of others.

Empathy is integral to the human condition. It's shaped the evolution of our brains and our impressive mental abilities.[5] Predating language,[6] empathy allows us to communicate effectively, and to bond with others, because if you're experiencing positive emotions via someone else, you want to be around that person. 'Good Sense of Humour' is found on every dating profile for a reason. And while the ability to detect and share the emotions of others may sound like something from science fiction, empathy is achieved via an exquisitely sophisticated network of neurological regions spread throughout key areas of the brain.

The key function of this network is 'action representation',[7] where the brain creates a representation of a specific action. This information is used when executing a corresponding voluntary movement, to guide and influence it. Action representation can occur when we think of specific movements, but it's especially important when we observe movements performed by others, because it allows us to *imitate* them.

That all sounds a bit technical, so look at it this way: you know when Sherlock Holmes gathers all the subtle clues from a crime scene (a fingernail, a spent match, a thread from a sweater, etc.), then works out in his head exactly what happened and who was involved, thus cracking the case? Action representation is the neurological version of that. Your brain accumulates all the sensory cues from observing someone performing an action, puts them all together into one coherent whole, works out what it means/represents, then figures out how it's done.

Imitation is a big part of how we learn and develop,[8] so we often imitate the action our brains have just observed and worked out. That's where the Sherlock Holmes analogy breaks down: repeating the crime he's just solved would be somewhat self-defeating. Simply put, 'action representation' is the process where our brains recognise what an action is, what it means, and how to do it.

To take an example from the early days of the human brain, imagine you're a primitive *Homo sapiens*, and you see a tribemate using a rock to crack open a coconut. Here's what we think is occurring inside your brain when you observe this 'action'.[9]

Firstly, the visual information our brain obtains from watching our fellow human attempting to crack a coconut is relayed to the superior temporal cortex, a key region for visual spatial awareness, and for integrating an egocentric and object-centred viewpoint.[10] Put more simply, it visually works out where things are in relation to us, and what they're 'doing'.

By doing this, the superior temporal cortex essentially creates a useful 'copy' of what we've just seen. It's sort of like scanning a photograph and saving it on your hard drive: it provides a version of the information that's easier to work with and utilise.

This information is then sent to mirror neurons (discussed earlier) in the parietal lobe (the upper-middle area of the cortex). Specifically, in the posterior parietal cortex. This brain bit has many functions, including combining sensory and motor activity, and forming *intentions*.[11] In particular, it recognises and encodes the actual movement(s) being observed, and which part of the body goes where (e.g. the slow lifting then rapid descent of the arm holding a rock). It also, crucially, extrapolates how *we'd* perform the same movement with our own body, and provides impetus to do so.

This information is then sent to mirror neurons in the inferior frontal cortex, another region with many important roles,[12] this time located at the front of the brain. Its main contribution to action representation is predicting the outcome, or 'goal', of the visible action. When watching our rock-wielding comrade for the first time, we'd likely think, 'Ah, they're doing that to break open the coconut, to get at the tasty bit'. This reasonable conclusion comes via the inferior frontal cortex, after it's fed information from the previous two regions. The inferior frontal cortex figures out what the *purpose* of the observed action is, and whether it's *worth* imitating.

So, when watching someone perform an action, i.e. hitting a coconut with a rock, our brain works out how to imitate it ourselves, by figuring out the 'what', the 'how', and the 'why' of it. Our superior temporal cortex assembles a neurological representation of the observed action, providing the 'what'. The posterior parietal cortex extrapolates the physical movements needed to copy it, providing the 'how'. Finally, our inferior frontal cortex works out the ultimate point of it, and whether it's worth imitating, providing the 'why'.

But then this information is relayed *back* to the superior temporal cortex area, where this whole process started.* Remember, this is the part that deciphers the action we're currently looking at. By looping the process back to here, our brain can compare

---

* In truth, all the described regions responsible for this imitation process are found in the *right* brain hemisphere. Their left-hemisphere counterparts are more involved with language and conscious communication.

what it's worked out and predicted about the action with what it is *actually* observing. Our brains basically go, 'Here's what I think will happen because of this action, here's what *is* happening . . . do they match?'

If they do, it means our action representation network has worked things out correctly, so the action can be imitated. In this example, if it deduces that the rock wielder's actions are intended to open the coconut, seeing that exact outcome will mean we've learned a new way to open coconuts. This is incredibly important: it means we can acquire useful new skills, *without* having to go through the laborious (and often hazardous) process of trial and error.

Of course, if the predicted and actual outcome *don't* match, imitation need not occur. If we see the rock wielder accidentally bring it down on their exposed other hand, resulting in injury and screaming, this wouldn't match the expected outcome our brain has produced. Thankfully, the inferior frontal cortex handles action *inhibition* too,[13] so can put the brakes on any impetus to imitate.

Overall, this diverse network of mirror-neuron-infused regions allows our brains to observe an action, and say, 'What is this, what does it mean, and how do I do it?', very rapidly and without much conscious input. It's a fundamental part of human learning and development.[14]

To bring it back to the original point, this network also plays a vital role in empathy. The link between action representation/imitation and empathy occurs via the insula, another region deep within the central areas of the brain, with a wide range of functions and applications, many of which are strongly associated with emotion. For instance, the insula is a key brain region for the experience of disgust.[15]

One particular part of the insula, the dysgranular field, is highly connected with the posterior parietal, inferior frontal, and superior temporal cortices,[16] aka our new friends which form the network responsible for action representation and imitation. And, thanks to the insula's many roles in emotion, the dysgranular field is also heavily linked to the limbic regions, extensively and heavily implicated in emotional processes.

Put as simply as possible, the dysgranular field acts as a hub which connects the neurological regions responsible for imitation, and those for emotion. The result of this is that, as well as the physical actions they're performing, the *emotional* signals being given out by someone we're observing* can also be deciphered, understood, and imitated (i.e. shared) by our brains.[17] And so, we get empathy.

Variations in the connectivity, influence, and activity of these vital neurological circuits may explain why the ability to empathise varies considerably from person to person.[18] But in general, it means empathy is a rapid, persistent, and largely subconscious process; we don't need to learn it, we can just do it.

This isn't to say our ability to empathise is fixed; we can develop, refine, and enhance it, via learning and experience.[19] However, that we're born with the ability is hard to deny, especially since even babies can do it.

Babies regularly display sensitivity to an adult's emotional state.[20] They even react differently to hearing other babies cry, compared to recordings of their own cries,[21] demonstrating an awareness of emotions that are *not their own*. And they often respond in kind, crying when they hear others like them cry. Clearly, some form of empathy is present in most human brains from day one.†

Further emphasising the link between empathy, imitation, and physical cues is our unconscious tendency to mimic the subtle

* And not just in the visual sense. The superior temporal cortex also contains the auditory cortex, the brain region where sound is processed. The emotional impact of smell suggests the olfactory system gets involved too.

† The matter of how neurodivergent people, like autistic individuals, process emotions and empathy is a more complex matter, which will be looked at later.

mannerisms and movements of those we're communicating with. Ever found yourself folding your arms when talking to someone who's doing the same? Or leaning the way they're leaning? This is what happens when the imitation systems of our brain are left 'unsupervised', because we're engaged with what's going on in the interaction.

This odd tendency exists for a reason, though. People apparently experience a positive emotional response to being mimicked, which they associate with the person who mimicked them.[22] As a result, they tend to behave more positively towards them. In fact, those who've been mimicked tend to behave more positively to others, too. They show more prosocial behaviour, to give it the technical term.[23] Indeed, individuals with greater recorded levels of empathy tend to unconsciously mimic others more often, leading to greater bonding and prosocial behaviour. This has been dubbed 'the chameleon effect'.[24]

Mimicry can happen voluntarily, as in consciously, too. Deliberately mimicking someone is a recognised way of gaining someone's trust, often resulting in them being more honest with you.[25] Then again, psychopaths often exploit this to manipulate others,[26] so it's not all good. But in general, this whole process – reading someone's emotions and movement and demonstrating them in turn – is largely an unconscious one.

This can have some surreal outcomes. While talking to someone, our respective (subconscious) brains and bodies are, in many ways, engaged in a distinct dialogue of their own, which can shape our feelings on a far more immediate, profound, *emotional* level. Maybe this explains why some people seem to 'click', in the romantic sense, despite apparently having little in common. They could be polar opposites on the intellectual or ideological level, but, beneath that, their emotions and bodies could be very in sync, communicating a strong emotional connection without having to involve the fussy higher brain regions. It's the default plot of every romantic comedy, but it's seemingly common enough in real life too – and this might explain why.

It also explains why I failed so spectacularly at hiding my grief from my family: my cognitive mind wanted to keep it from them, but my subconscious emotional brain had no time for any of that nonsense.

But empathy is not always a positive thing. It also means we can share someone's discomfort, anguish, and suffering. Indeed, the fundamental nature of empathy is regularly demonstrated by the recognition and sharing of *pain*.[27]

If you tell someone about when, as a child, you were hit in the mouth by a swing and ended up biting through your tongue, they'll likely be horrified, visibly wince, perhaps even clamp a hand over their own mouth.* This is because we're particularly sensitive to other people's pain; if we see/hear about someone's grim injury, we tend to recoil as if it's happening to us. Even though it clearly isn't, and logically cannot.

The phrase 'I feel your pain' isn't just a cliché; when we observe or hear about someone experiencing pain in a particular part of their body, the sensorimotor areas of our brains show raised activity in regions that correspond to the same locations on our body.[28] In layman's terms, if we see a spike going into someone's left foot, our brains show activity suggestive of similar happening to *our* left foot. It's no wonder we react to other people's pain as if we were experiencing it; to an extent, we are.

Obviously, our own pain is much less than that of the person we're empathising with. This makes evolutionary sense: back in the distant past, experiencing the exact same pain as an injured person

* This observation is based on how people have reacted when I tell them about the time I bit through my tongue as a child after being hit by a swing.

would have meant a predator could bite one single human and incapacitate the whole tribe, as they all writhe around in empathy-induced agony. And that's not a good survival strategy.

But here's the interesting thing: we don't *need* to feel someone's pain to empathise with it.

Due to some fluke of genetics, biochemistry, or neurology, a small number of people don't experience much pain, or any pain at all, in response to injury.[29] If you show such people footage of someone else being hurt, then ask them to accurately guess how painful that person finds it, they're really bad at it; they've no similar experiences to compare it to.

However, despite this, they're as good as anyone at guessing how much pain someone is in if they see that person's *emotional* reaction: their facial expression, movements, the noises they make, and so on. They can still empathise with someone else's emotional distress.[30] And this is important, as it demonstrates that we can have a shared emotional state even in the absence of a shared sensory experience.

Similarly, we don't even need to be able to form emotional facial expressions ourselves in order to recognise them in others. People with Moebius syndrome, a rare form of neurodegenerative disorder which causes facial paralysis from birth, have little to no difficulty recognising other people's facial expressions.[31]

This shows us that empathy isn't just some kind of evolutionary by-product of ego or self-preservation. It's a distinct process, an ingrained, fundamental aspect of human nature, and the emotions we experience.

What's the purpose of empathy, though? What advantage does it give us? Given how empathy, in some form, seems to have been around in nature for far longer than the human race,[32] it must be a very useful ability.

Many argue it's to do with altruism, the selfless concern for the wellbeing of others. When you can share other people's emotions, you're obviously going to be far more invested in their emotional state and wellbeing, because it directly affects you. Despite the many who insist that life is fundamentally 'dog eat dog', 'every man for himself', 'survival of the fittest', and all that, there's much evidence to suggest that we humans (and other social species) are hard-wired to be cooperative, communicative, and altruistic, right down to the genetic level.[33]

Unfortunately, and counterintuitively, many argue that our altruistic tendencies are, in fact, *selfish*.[34] This is less of a contradiction than it sounds, because our altruistic tendencies are typically reserved for our kin: in this case, those we are related to, or otherwise emotionally bonded with.* We spend practically all our time around such people. And, if we make sacrifices for them, they, having similar tendencies, will be far more inclined to return the favour, and do things for us when the opportunity arises. Essentially, it's argued that our altruistic tendencies are a form of emotional investment in those we have relationships with. An investment that we expect to be repaid, possibly with interest.

Also, possibly thanks to this process, when we encounter someone from outside our kin, someone we've no existing bonds with, we can be far more wary of them, possibly to a hostile degree. We've no emotional investment in them, so have no reason to trust them. It's argued that our altruistic tendencies actually *reinforce* an instinctive xenophobia.[35]

Empathy can seem selfish on a more individual level, too. If someone's happy, and you're sharing their happiness, then you're motivated to enhance their happiness, because that makes you happier too. Similarly, if someone's sad or experiencing some other

* In contrast to most species, the powerful human brain allows us to value and prioritise people we aren't genetically linked to/mating with. We can grasp the concept of friends, colleagues, teammates, etc.

negative emotion, you're motivated to help resolve their unpleasant emotional state, so as to improve your own.[36] Overall, it can seem that the very concept of human selflessness is fundamentally *selfish*. Something both confusing and depressing.

However, don't write off humanity just yet because, despite all the logical counterarguments, a lot of research suggests that we *do* behave altruistically because we're more concerned for the other person's wellbeing, not our own. For instance, studies show that individuals who help someone tend to retain an interest in that person's wellbeing long after the event, even if they don't know them, and even if they know their efforts to help them won't have any lasting effect.[37]

The implications are profound: it suggests we retain an instinctive drive to assist others and a concern for their wellbeing, even if there's zero chance of them returning the favour, and even if our efforts to help them ultimately failed. There are no positive emotions to share, or emotional debt to be claimed. Even if there's nothing to be gained from improving the wellbeing of others, we often *do it anyway*.

This suggests that empathy makes us care about people because . . . well, just because we care, I guess? Empathy makes it easier to care about others, maybe even compels us to do so. And that's just for people we have no existing connection to. Imagine what people would do for someone they already cared deeply about?

Luckily for me, I don't need to imagine this. My wife and children demonstrated it repeatedly, during the most difficult time of my life. I'll never forget that.

Feelings contagious: how we get consumed by the emotions of others

Have you ever walked into a room and suddenly felt tense and awkward due to the 'frosty' atmosphere? Usually, it's because those already in the room have had a big argument before you showed up. However, you have no way of knowing that; you weren't there, and nobody has told you anything. Those already present may even be trying to talk and act normally. Nonetheless, you still know that something has happened, because you can *feel* it, i.e. you have an emotional reaction. But you don't know why, or where it came from.

I keenly felt this myself, at my father's funeral. It was, predictably, a bleak and melancholy occasion, and made me even sadder than I already was. But even here, my rational brain was still buzzing away in the background, wondering *why* I was sadder. Obviously, I was emotionally devastated by my father's death, but the funeral happened nearly two weeks later, so this wasn't a 'new' thing. And I'm not particularly religious or spiritual, so that side of things shouldn't have bothered me.

Ultimately, funerals just *feel* sad, because they're full of sad people, and that affects us. Even if you don't know the deceased and are just there to support someone who did.

Why does this happen? A logical assumption would be that this is just another example of empathy. But this assumption is wrong. In situations like wandering into a room that 'feels' tense, can you pinpoint the specific person you're empathising with? You don't know what was said, who was in the right, or who was out of order. It's more a general emotional 'vibe' or 'atmosphere'. And here's the rub: if you can't say exactly who you're empathising with, it's technically *not* empathy.

We've seen that the neurological mechanism that gives rise to empathy is dependent on the brain's ability to recognise what someone else is doing, work out what it means, and figure out how we'd do it ourselves.[38] The key element there is 'someone else'. We recognise that another individual is performing the action, which means that, while it's largely a subconscious emotional process, empathy still involves cognition; you need to be consciously aware of another individual doing or feeling things which are distinct from your own actions and emotions.

The ability to recognise and understand that other people are individuals with their own internal mental states, which may differ from ours, is something very few species besides us humans are capable of. It's a major cognitive achievement, and one that we've seemingly evolved dedicated brain regions to facilitate. Specifically, our brains include a region around the paracingulate sulcus, another part of the prefrontal cortex, that plays a prominent role in attribution of *intention*.[39] Basically, it seems we've a specific brain region for figuring out why other people do what they do.

This region contains many spindle cells, a class of neurons with long projections which link multiple different regions of the brain together.[40] They're seemingly involved in coordinating widespread activity involved in both emotion and cognition, which would be a very useful feature for figuring out what someone else is thinking and feeling.

These spindle cell neurons are, thus far, only found in great apes and humans.[41] The implication is that, for us and our smarter primate cousins, knowing how others are thinking and feeling, and differentiating their feelings from our own, was a significant evolutionary advantage. Yet another thing which suggests emotions shaped us humans to be the way we are.

So, empathy requires conscious awareness that the emotions you're experiencing didn't originate within your own mind,[42] but from another person. However, here's the thing: we can still detect and experience 'external' emotions, without consciously recognising that they come from someone else. It's just that, if this happens, it's not empathy. It's emotional contagion.[43] And while the two inevitably have a great deal of overlap in the brain, they also have important distinctions.

Emotional contagion is a more primitive form of – or maybe even a component of – empathy.[44] It's a means of sharing the emotions of others, but lacks the vital element of self–other distinction, without which we can't experience empathy. In fact, you may recognise the term emotional contagion from the discussion about music in Chapter 3. While we can certainly perceive that music has certain emotional qualities, we can't recognise what music is 'feeling', because it's not feeling anything. Ultimately, it's just sounds. Therefore, we can't empathise with music. But we can still be emotionally affected by it, because emotional contagion *is* possible with music.* And that's the key difference: with empathy, you know whose emotions you're sharing; with emotional contagion, you don't.

In the example of walking into a frosty room, the post-argument individuals' roiling emotions are influencing their bodies and actions, despite their best efforts to the contrary. They're displaying signs of hostility, tension, animosity, etc. And your brain is picking all this up. Your mirror neurons are still being activated by what we perceive others as doing, and the emotional information they detect is being shunted to your limbic system, causing you to experience similar emotions yourself. But in this case, that information isn't being shared with your higher cognitive areas, so you experience a new emotion, but aren't consciously aware of why, or where it's coming from.

Clearly, this doesn't happen all the time. We don't automatically and unknowingly share the emotional state of every single person who wanders into our eyeline. Emotional contagion is most likely when we're exposed to many people who are all experiencing the

---

* Though it's worth pointing out that music featuring singing *would* allow for empathy, because the emotions can be attributed to a specific person or persons.

same strong* emotion, one affecting them in more easily detected ways. But multiple people displaying the same emotion means our brain can't pin the emotion it's detecting onto someone specific. So . . . it doesn't.

That's why we feel scared when we're part of a panicked group, even if we've no idea what they're all afraid of. Or sad at funerals, even if we don't know the person who died.

It may seem quite alarming that our emotions can be essentially imposed on us by others, who are themselves in a highly emotional state. Aren't we humans, with our mighty brains, meant to be smart, independent individuals? Sure, but when you see someone yawning, what do you do? Whether you want to or not, you yawn too. Because it's undeniable that yawning is contagious. And not just for us humans. Yawning is contagious in many other species too, like chimps, dogs, and more. It can even occur *across* species: if we see a dog yawn, we're often compelled to do the same, even though we have a completely different jaw. Indeed, efforts to stifle the impulse to yawn often make it *stronger*.[45] It's a very powerful reflex.

The reason we yawn, and why it's so contagious, remains unknown, as do the neurological mechanisms responsible for it. However, recent studies suggest that yawning may be to fatigue what laughter is to amusement: a way of communicating your internal state to others.[46] Knowing that one of your group is very tired, and therefore compromised, is important information for a social species relying on each other for survival. Communicating and responding to this, rapidly and reflexively, would be a very useful trait. Simply put, yawning is an involuntary body movement, triggered by our internal state, that displays how we're feeling to others, which compels them to do *and feel* the same. And it all happens without our cognitive brain regions getting involved.

* As in, a very arousing one, one of high salience.

The example of yawning shows us two things. Firstly, that it's perfectly normal for other people to have a powerful yet sub-conscious effect on what we do and how we feel, so emotional contagion is no great leap. Secondly, like yawning, emotional contagion evolved for good and useful reasons. Our brains wouldn't do these things, readily and regularly, if they hadn't repeatedly proven to be useful. Emotional contagion allows us to quickly tune in to what's going on around us via the feelings of others, without the delay that would be introduced by having to painstakingly figure it out via our complex cognition.

Instinctively laughing or rejoicing with the happy people around you can help you bond with them, a priority for humans.[47] Automatically becoming scared and agitated when amongst others who are in such a state makes you ready for dealing with what-ever caused their fear and anxiety, whereas logically working out what the threat may be takes effort, and time, during which it may decide to leap out from the shadows and eat you.

But emotional contagion has a dark side, too. Few would deny that when people gather in large enough groups, they often behave and think in less rational and reasonable ways than they would individually. And while our natural tendency might be to instinc-tively care for others, there are countless real-world examples of emotional contagion making us more hostile, destructive, and aggressive to others who have done us no wrong. How, and why, does this happen?

Especially potent emotions can hamper our abilities to focus on and logically assess things.[48] When we're deliriously happy we think little of paying for things we can't really afford; if we're consumed by terror we will recoil from even the most innocuous or innocent stimulation; and when we're utterly furious we can be genuinely dangerous to be around, because we have so little control over our actions. The problem is, while there's often a useful, but delicate, interplay between the cognitive and emotional outputs of our brains, when emotions get *too* powerful, it can throw a spanner in the works, and cognition (and therefore our cognitive ability to control our emotions) is compromised.

This doesn't necessarily mean that cognition and emotion are fundamentally different things that work against each other. It could just be due to the limitations of the underlying physics of the brain. When our brain is doing something, we're using, 'activating', the parts of it responsible for this process. But despite their impressiveness, our brains are still biological organs. When an area of the brain is activated, it uses more energy. Therefore, it needs more biological resources, more fuel – namely glucose and oxygen, the stuff that cells run on.

As with any other organ, these are supplied to the brain via the blood supply. However, because the brain is composed of tightly packed, delicate, and metabolically demanding neural tissue, there's little space for blood vessels. One unhelpful outcome of this is that there's limited flexibility in the brain's blood supply, so redirecting it, and the vital resources it conveys, to where it's particularly needed, is quite challenging.[49]

It's like the brain is a restaurant with a hundred tables, all occupied by customers. The waiters working at the restaurant are the brain's blood supply. Unfortunately, there are only five of them, meaning only five tables can be dealt with at any one time. If a sixth table suddenly needs attention, then either it must be ignored, or one of the current tables being tended needs to be abandoned.

This means it's not metabolically possible to 'activate' all of the brain at once. That's why it's really hard to compose a song while reading a book, or perform complex mental calculations while engaged in a detailed conversation, and why there are literal laws about doing things other than driving whilst driving.[50]

Given how, despite their many overlaps, logical thinking and intense emotional experiences are often supported by distinct neural regions,[51] they'd presumably compete for allocation of the limited resources the brain has thanks to its constrained blood supply. Indeed, studies have shown that strong emotions lead to a predictable increase of neurological activity in associated areas of the brain (like the amygdala), alongside a corresponding *decrease* in important cognitive regions, like the dorsolateral prefrontal cortex.[52]

This may explain why emotion and cognition peacefully co-operate most of the time, but when emotions get too powerful and start hogging the brain's resources, our cognitive abilities are compromised. They have to do more, with less.

This is problematic enough already, but when you add emotional contagion into the mix, it gets serious, because your brain is picking up the emotions from the people around you, who are reinforcing, even amplifying, your own emotions. This would explain why emotional contagion can lead to the famous 'mob mentality', where you effectively lose your self-awareness and self-control when you're part of a highly emotional group, meaning you think and act in ways you never normally would.

The exact mechanisms and processes of this phenomenon – known as deindividuation[53] – are still the subject of much debate. But one thing nobody disputes is that emotions play a prominent role in deindividuation. Which makes sense, because strong emotions seemingly disrupt our ability to think rationally.

More specifically, studies suggest that the anterior prefrontal cortex is responsible for evaluation of self-generated responses, and is involved with actions that require monitoring info that's been internally generated.[54] To translate: whenever we do or think anything, the anterior prefrontal cortex goes, 'Why did I do that?', 'Where did that come from?', 'Was that a good idea?', 'Should I do that again?', and so on. This process is a big component of how we refine and control our emotions.[55] So, how would we act if this ability were reduced, or shut down? More unpredictably, more impulsively, with much less self-awareness and self-control: just like we do when caught up in a mob.

Mob mentality becomes all the more powerful when there's an external threat or rival group to focus on, which provides an easily recognised external focus (or 'target'), thus improving and maintaining unity and cohesion for the mob itself.[56] Think of rival football fans clashing violently, rioters charging at lines of police, or even the angry villagers pursuing a supposed monster with pitchforks and torches, a staple of classic horror. In these instances, people become so emotionally stimulated by those around them that they lose their ability to recognise and understand the thoughts and feelings of individuals 'on the other side'.[57] Ironically, this would suggest that emotional contagion can, in extreme cases, make us *incapable* of empathy. Is it any wonder that mobs can be so dangerous?

This isn't to say that emotional contagion is a bad thing, because clearly it often isn't. But it helps explain why we reflexively feel so sad, or happy, or angry, or scared, when in situations where we're amongst people who are also feeling (and displaying) those emotions. Be they situations where an argument has just happened, amidst an angry mob, or attending a heart-breaking funeral.

No wonder our emotional experiences can so often be so weird and confusing: sometimes, they're not actually *our* emotions at all.

It makes you wonder: how do we get anything done?

Emotional labour: emotions in the workplace

Work gets a bad rap. We call it 'the daily grind', describe ourselves as 'living for the weekend', espouse the merits of a 'work–life balance', and so on. Why, though? There are many things in life that can result in very negative emotional reactions, but we rarely consider them to be negative by default, like how we so often do with the world of work. What's going on there?

Many psychologists, career coaches, and self-help authors have attempted to answer this question before me, but after my father's death I started to engage with it in a new light. Could it be that my own emotional confusion during the grieving process might be due, at least in part, to my employment history? A strange proposition, I know, but let me explain.

Before my neuroscience doctorate, I worked as an anatomy technician for a medical school, where I embalmed cadavers. People would agree to leave their body to the university upon their death, and it was my job to chemically prepare and process said body, so medical students could safely use it for learning anatomy and practising surgical techniques. Basically, for roughly two years, my day job involved embalming and cutting up the dead bodies of the recently deceased. It was as pleasant as it sounds.

It also had lasting effects on me: I've still got an extremely strong stomach for anything involving blood or surgery, and I've never lost a drunken argument about who's had the worst job. But, as I struggled to deal with my feelings in the wake of losing a parent, I began wondering if my grim old occupation hadn't also altered my emotions, in less than helpful ways.

It's emotionally challenging, engaging with a dead body. Every year, a few new medical students promptly dropped out of the programme because they couldn't cope with doing so.* Unfortunately, as an employee I didn't have that option. I just had to deal with any emotional discomfort I felt and carry on. I did this by becoming

* Just getting into medical school takes many years of hard study and achievement. Sacrificing all that, rather than face the dissection room, shows just how powerful an emotional reaction dead bodies can provoke in even the most intelligent, driven people.

as emotionally disengaged as possible. I regularly utilised my cognitive brain's ability to rein in or restrict my emotions, by persistently convincing myself that, despite appearances, the cadavers I handled weren't 'people', merely inert objects. Not the nicest approach, but it worked.

Did I overdo it, though? It's like, if you stretch out the elasticated waistband on your pyjama trousers, it'll snap back into place. But if you pull it too far, too hard, or for too long, the waistband will be stretched beyond its elastic limit. It won't snap back, it'll go all floppy, and your pyjamas won't provide the support needed. I was worried that this had happened to me, only instead of pyjamas, it was my brain's emotional processes, which I'd argue are more important. And if this had happened to me, could it also happen to anyone else, regardless of how grisly or not their working lives might be?

To answer this, the first thing to consider is how the very nature of work causes us to experience things that rarely occur elsewhere. For instance, imagine one of your friends sitting you down and listing every mistake you'd made over the previous year, before explaining how you needed to do better to maintain the friendship. This would be a very emotional experience; you'd be humiliated, upset, and furious with your soon-to-be-former friend.

Thankfully, friends don't do this to each other. But it's very common in the modern workplace, as anyone who's experienced an annual appraisal or performance review will know. And the unpleasant emotions this grim ritual induces in us don't just vanish into the ether once it's done. They can have lasting, fundamental effects on us, thanks to a neurological mechanism known, appropriately enough, as 'appraisal theory'.[58]

Appraisal theory began as an attempt to explain why people regularly have different emotional reactions to the same things. You may get teary-eyed at a period drama, while the person sat beside you is bored senseless. Some love skydiving, others panic at the very notion. If our emotions were created by the same hard-wired mechanisms in each brain, as many have argued, then logically we should all have the same fears, likes, dislikes, and so on. That doesn't happen, though. Appraisal theory is one possible explanation as to why. It argues that our emotional reactions are actually the result of our brain experiencing something, *appraising* what this something is and what it means for us, and using the results of this appraisal to figure out the appropriate emotional response.

Say you see a large shaggy dog bounding towards you. Your brain may go, 'Big dog approaching. I like dogs. This one looks playful, so the appropriate emotions here are happiness and excitement.' Alternatively, your brain could go, 'Big dog approaching. A dog bit me when I was small. I don't like dogs, and there's a large one incoming. The best emotion here is fear.' Both are examples of your brain appraising the situation and working out the relevant emotional response. But which is correct?

Both. Both are completely valid reactions, even though they produce radically different emotional responses. Our brain's appraisals are based on memories, understandings, and assumptions that vary considerably, from person to person. In a nutshell, it's not what we're experiencing that determines our emotional reaction, but *how our brain interprets it*. And this interpretation varies considerably between individuals.[59]

Where this gets interesting is that the memories and past experiences that influence our appraisals *themselves include emotion*. Therefore, the emotions we recall from our pasts influence the ones we experience now. Accordingly, many scientists differentiate between primary and secondary appraisal. Primary appraisal concerns our initial emotional reaction: someone criticises you, your brain appraises this and concludes it's some form of personal attack, so you get angry. This is the emotion resulting from the primary appraisal.

Secondary appraisal is where you appraise the *results* of your primary emotional reaction and incorporate this into any future appraisals. In this case, say the anger you felt after being criticised motivated you to retaliate, and you attacked the criticiser. Unfortunately, the criticiser was your boss, during an important meeting. And now, thanks to the primary emotional reaction, you're unemployed.

Here, performing a secondary appraisal allows your brain to learn that the results of the primary appraisal had negative consequences. Admittedly, it's typically faster and subtler than this exaggerated example, but the outcome is basically the same: next time you have a similar experience (i.e. someone criticises you), the cognitive appraisal, and the resultant emotions, will be based on more (and hopefully better) information. This ideally results in you having more beneficial emotional reactions.

This mechanism would help explain how work can have many a profound and lasting effect on our emotions, but it isn't exclusive to work; it applies to any novel emotional experience, in any context. It isn't necessarily *bad*, either: dealing with unique and unfamiliar emotional experiences helps expand your emotional understanding and abilities, which enhances our emotional competence.[60] Indeed, secondary appraisal is a big part of our coping abilities, of how we learn to deal with things like stress. Interestingly, much of the data supporting this argument comes from psychological studies of the workplace.[61]

As well as my own morbid occupation, there are also numerous jobs where keeping your emotions in check is helpful, even vital. A teacher who screamed loudly in frustration when faced with an unruly classroom wouldn't be teaching for very long. A paramedic fainting in horror at the sight of blood is of little use to those whose lives may depend on them. If you work handling dangerous substances like toxic waste, a tendency to panic and get all twitchy when near anything dangerous is a genuine liability.

However, if controlling your emotions is a necessary part of your job, the tendency to do so, learned over time, isn't something you leave behind when you clock out and go home for the day. If you know a nurse or paramedic, you'll know how unflappable they can be. And it's often easy to tell if someone you meet is a teacher, as they tend to give off a weird authoritative 'aura'.* This is because, while it's fine to say people should maintain a good 'work–life balance', the truth is we use the exact same brain for both, so what that brain does in work will inevitably impact what it does when we're not working.

This comes back to our common perception of work as a more negative emotional experience. Even if you love your job, there will regularly be days when you resent it, find it gruelling, or simply don't want to do it. And I'd confidently state that people who love their job are very much a minority of the world's workers. Indeed, the word 'burnout' is used often these days, and for good reason.[62] Defined as 'a state of emotional, physical, and mental exhaustion caused by excessive and prolonged stress' which usually means you can no longer function normally, a big factor in most cases of burnout is an excess of negative emotions in the workplace.[63] Why would work be a reliable source of negative emotions, often to the point where we can no longer handle them?

Part of this can be blamed on how, in most lines of work, emotions aren't really considered. If your job involves compiling spreadsheets, or building walls, the emotions spreadsheets or brick walls cause you to feel are typically irrelevant. Whether you do your job with a smile, or a scowl, is often of no consequence. The problem is, you'll keep experiencing emotions at work regardless.

---

* Many stand-up comedians I've encountered are former teachers. The ability to control a room full of rowdy individuals, who may have zero interest in what you're saying, is a valuable transferable skill in this case.

Even if your dreadful boss honestly thinks employees are mindless peons, that doesn't make it so.

In truth, work inflicts a lot of stress on us, often in ways the human brain is particularly sensitive to. Loss of autonomy[64] (e.g. being micromanaged), loss of social status[65] (the obligation to be subservient to a rude, obnoxious customer), wasted effort[66] (a project you've worked on for months being scrapped to save money): all these things, and more, are everyday experiences in many workplaces. All are things our brain doesn't enjoy, so reliably induce negative emotions.

Unfortunately, most jobs don't give you the opportunity to process or deal with these emotions in healthy ways. You can't punch a wall when someone angers you, you can't wander off and have a good scream or cry, you can't respond in kind when belittled by a horrible customer or toxic boss. Your negative emotions just build up in the brain, unprocessed,[67] like exhaust fumes leaking into the car that you're driving.

Thankfully, many modern workplaces have started taking employee emotions into account. Some offer resilience training, where workers receive coaching on how to better handle stress and negative emotions, which can have very positive effects on mental health and wellbeing.[68] There's also a potent modern trend of employers emphasising the importance of employee happiness.[69] What's that, if not a clear acknowledgement of the emotions of the workforce?

Yet burnout, workplace stress, and poor employee satisfaction are still enormous problems, with record numbers of workers being disengaged and unhappy with their jobs.[70] One possible explanation is that, whatever their intentions, the emotional wellbeing of employees typically isn't the top priority for those in charge of companies or organisations. If it's a business of any sort, profit, the so-called 'bottom line', is typically the number one priority.

Workers are, ultimately, a means to that end.

This isn't to say that healthy profits and healthy employees don't overlap. If workers find a job debilitatingly stressful, they'll be unable to do it, so addressing the emotional burdens of the work makes financial sense. However, employers and bosses are known to often adopt a cynical approach to this. Resilience training, mindfulness workshops, etc. can be very helpful, but they don't do anything to tackle or address problems like unrealistic workloads, long hours, poor pay, and so on. Also, according to the testimony of countless people, many employers seem to treat this type of training as a blank cheque, assuming it gives them free rein to dump ever more work and demands on workers, even if that's what harmed their emotional wellbeing in the first place. It's fine. Because they're resilient now. Right?

Except, it's not fine. If anything, it adds even more stress to the individual worker, making it *their* responsibility to learn how to cope with the negative emotional consequences of their taxing job, something which takes time and effort, which are often in short enough supply as it is.

Then there's the efforts to make employees happy. This usually stems from 'the happy-productive worker' thesis,[71] which shows that happier workers are more productive, i.e. they do more work, for no extra money. What employer wouldn't want that?

Predictably, it's not that simple. Another consequence of the appraisal theory of emotions is that making a large and varied group of people consistently happy – i.e. experiencing the same emotional reaction to the same thing – is virtually impossible. That's why employers' efforts can seem somewhat simplistic. Casual Fridays, employee-of-the-month schemes, team-building exercises, annual bonuses, in-chair back massages,* and so on: these

---

* Multiple people have assured me that this is a real thing.

approaches may make *some* employees *briefly* happy, but overall they're like trying to improve your car engine's performance by dousing the whole thing in oil and hoping for the best.

Also, people *don't like* being emotionally manipulated. There's much research confirming that humans instinctively dislike being told they have to do something, or having their options taken away.[72] Have you ever been told to 'cheer up' while in a bad mood? This unsolicited instruction often has the opposite effect. Because what right does anyone have to dictate *your* emotions?

So, in summary, work is more likely to cause us to experience negative emotions due to the very nature of what it asks of us. Couple this with the persistent efforts of employers to actively manipulate/impose the emotions of workers in ways that benefit the organisation, and is it any surprise that we tend to think negatively of work?

Where work differs from the innumerable other aspects of modern life that regularly cause us to experience negative emotions is that work actively encourages us to *suppress* these emotions. We can't help feeling them, but we are frequently unable to communicate and share them in the manner we have fundamentally evolved to do. Because of the rules and expectations of the workplace.

This is bad. Given how emotional processing and cognitive appraisal works, our brain can readily develop a habit of emotional suppression from the workplace. And given how important emotions are to us, this can have very unhelpful consequences. It can distort and disrupt your sleep, your mood, your home life.[73] It can put strain on important relationships.[74] It's even been linked to the onset of depression.* [75, 76] So, while there may be logical reasons

---

* Jobs requiring a significant degree of emotional suppression have been shown to make workers more prone to depression and anxiety. Retail and call centre jobs are the most frequently cited examples, as they regularly involve the need to remain civil while being harangued by belligerent customers.

to not express and communicate your emotions in work, it can be a harmful approach for individual workers in the long run.

But consider this: what if expressing and communicating emotions *is* what you do for work? While less common, such jobs do exist. The most obvious example is acting, a great deal of which boils down to portraying specific emotional states, while in character. Based on what we've seen so far, professional actors should be the most emotionally healthy and upbeat people around, because they regularly get to express themselves emotionally while at work.

Yet studies show that those in the acting and performance profession are *more* prone to issues like anxiety and depression when compared to the typical population, not less.[77] This didn't really match up with my predictions.

To find out what might be going on here, I spoke to actor, writer, performer, and hardcore Welsh person[*] Carys Eleri about how acting really works behind the scenes. Among her many credits, Carys played the lead role of Reverend Myfanwy Elfed in the celebrated S4C Welsh-language drama *Parch*, which ran from 2015 to 2018. And from what she told me, with acting, even if you're doing exactly the sort of work you want to be doing, it can still take a considerable emotional toll:

> The role [of Reverend Myfanwy] was every actor's dream. It was the lead part, it was meaty, it was diverse, it was interesting, it was dramatic. I was overjoyed when I got it. But actually *doing* it? God, that was hard work.
>
> I felt I had no actual life outside of the shoot – and I love life! It's a conundrum because you don't want to appear ungrateful.

---

[*] Most people consider me to be very Welsh, but Carys makes me look like someone who just visited Wales once for a weekend holiday.

I was over the moon to get a lead role in a juicy drama, but the woman on screen was really knackered, stressed, and unhappy most of the time outside of work, because on the 1–2 days off a week I had, I was so exhausted I could hardly get out of bed.

I asked Carys about the unique form of emotional labour required in her profession, performing heart-rending or emotionally brutal scenes, where she had to convey a range of powerful negative emotions, on command, for prolonged periods.

If you're filming scenes in a hospital, you may not have access to it for very long. So, you've got to do all the heart-breaking hospital scenes in one or two days and edit them into the right order later. We sometimes ended up filming for nine or ten hours straight, doing scenes where I had to cry or be distraught each time, or multiple times. It was exhausting. It drains you.

To make things harder for actors such as Carys, the human brain is, as we saw previously, extremely adept at recognising the emotions someone is expressing, via the many subtle and complex cues they give off when doing so. However, if those cues are missing or distorted, it instinctively strikes us as *wrong*, and we react negatively to it. Hence canned or false laughter is so grating, early CGI characters are creepy, and a bad actor is very easily recognised as such.

An important aspect of this is that many of the physical cues we display when feeling an emotion involve muscle movements and bodily reactions that we don't have conscious control over. We can't choose to have flushed cheeks, we can't willingly make our hair stand on end, and every wedding album the world over demonstrates how difficult it is to convincingly smile on command.

How does a decent actor get around this? Usually, by *genuinely* feeling the emotion they're meant to be portraying. Several actors have since told me that in drama school they're often made to experience and re-live painful experiences and memories, to improve their dramatic performance abilities. Presumably, this practice makes it easier to recall and invoke the emotional experience as and when needed for a character portrayal. Carys herself has relied on this approach a lot:

> I'm a very empathetic person anyway, but I've had so many friends who have experienced cancer and died, I lost my father to motor neuron disease, I've just gone through intense pain and grief with so many people. This does mean, though, that when I'm in these roles, playing these dramatic parts, I know how they should be played, how the character should be feeling as they deal with loss and heartbreak. Because most of the time, I've experienced it myself.

This distinction between real emotions and 'faked' emotions is really important, because it means actors regularly put themselves in a genuine negative emotional mindset as part of their role. This can be hard to break out of, meaning they stay angry, sad, or afraid, long after the scene has ended and everyone's gone home. Carys experienced this exact thing while filming the first season of *Parch*:

> For fourteen hours a day for four months I would convince myself I was dying! By the end of the shoot I was constantly convinced I was living with a husband I wasn't in love with anymore, had two kids, had some major hots for an undertaker while heading to an early grave. How relaxing!

It's a common problem for actors: they get caught up in a role, so the emotional turmoil they put themselves through for it doesn't just stop once the director yells 'that's a wrap'. But while for most people the impact of a negative emotional experience can eventually fade, as we process what we've gone through and the fading affect bias does its thing, the work of many an actor involves dredging negative emotional memories back up and reliving them all over again, via many different characters and performances, keeping those negative emotions fresh and active in the long term. That this is a known problem with PTSD[78] suggests that it's not the best thing for our mental wellbeing.

This issue has not gone unnoticed. The mental and emotional strain of acting has been recognised by many studies now,[79] and interventions and methods of alleviating the damage of this have been worked out and implemented in many areas of the industry.[80] For example, many actors are now encouraged to 'de-role' once they're done: to perform some ritual or regular gesture that marks a clear divide between the performance and reality, which allows them to leave their character (and the emotional demands of it) in the dressing room, so to speak.[81] Over many years on the show, Carys learned how to do this exact thing. The fact that she remains one of the most upbeat and friendly people you could ever hope to meet suggests her dramatic roles and experiences haven't done her lasting emotional damage.

And it's worth pointing out that acting isn't automatically all stress and negative emotions run rampant. There are many ways in which acting and performing can be *good* for your emotional and mental wellbeing. Drama therapy is an established and helpful practice,[82] after all, allowing people to express and work through their emotional neuroses and problems, in safe and controllable ways. This presumably helps in the same way as listening to sad music helps us emotionally process such feelings.

This can help actors in their normal working life too. Carys still remembers a very talented co-star, who was going through an incredibly hard time emotionally in her personal life, but still seemed perfectly cheerful and composed when they weren't filming:

> As soon as the cameras rolled – she was in that moment, her emotions were clearly bubbling on the surface and deep within her delivery of each word. I couldn't help but react – tears running down my face too. I asked her 'how do you do that, take after take?' She said 'This is therapy. Right now, I'm in so much pain and I have to keep it together, to do everything I do. With this scene, it allows it all to tumble out of me.' I probably knew deep down that this was the case, hence me crying, not just at a woman acting a script I'd read and studied over and over, but a woman who was carrying so much pain, stress, and responsibility, whose only moment to truly engage with her emotion was here, through the art that we were collectively making. Through creativity.

Maybe the real issue is that work doesn't just cause us to suppress our emotions, it often compels us to communicate the *wrong* emotions. While at work, we so often must maintain a front which is at odds with what we're really feeling. We laugh at our boss's terrible jokes, nod and smile at the belligerent customers, act calm and confident when faced with a daunting task or impossible deadline. All the while, our inner emotional state is completely different to the one we're communicating.

That may be why actors being 'allowed' to communicate emotions doesn't protect them from the emotional toll of work. They're often expressing and displaying emotions they wouldn't, or shouldn't, be feeling if they weren't working. And because our brains are always whirring away, observing and learning and tweaking our emotional processes based on what happens when we communicate our emotions to the outside world, the distorting emotional effects of work can build up, confuse matters, and impact on our overall wellbeing.

This doesn't have to be the case; many jobs and workplaces are introducing measures that are genuinely mindful and considerate of the emotional wellbeing of those who work there. But this seems to be a relatively modern phenomenon, and it's not the default norm yet.

Maybe one day we'll reach a situation where we're all able and permitted to accurately express our emotions at work. But to make that the norm? It's going to take a lot of, well, work.

It's another interesting point, though, the fact that some workers are allowed to express their emotions, and others aren't. It reveals that, in the world of emotional communication, not everyone is equal. And the deeper you delve into the data, the truer this becomes.

Emotional exclusivity: who do we empathise with, and who do we not?

Even though I was physically cut off from practically everyone in the aftermath of my father's death thanks to the pandemic, modern technology meant I was still readily contactable. So, I got many messages from people expressing profound sympathy, love and support, offers of help, and more. It was all very emotional. Predictably, this made me emotional in turn.

Unfortunately, the emotion I responded with was anger. These messages often *enraged* me. And my reactions weren't pretty. Being told, 'I wish there was something I could do to help'? Well, there isn't. We're under lockdown, and you live hundreds of miles away. This message exists purely to make *you* feel better for saying it. Screw you! 'I'm so sorry for your loss'? Why? Are you responsible? If not, you're just wasting my time and mental energies when I have little enough to spare as it is.

Before going any further, let me clarify that everyone who messaged had 100 per cent good intentions; my reactions were completely unfair, unwarranted, and unrealistic. Thankfully, I never *said* any of those things; I just thought them. That's the headspace I was in. In my defence, anger is a common aspect of grief.[83] The loss of someone close to you, particularly in tragic circumstances, feels catastrophically unfair, and perceived unfairness provokes anger in the fundamental levels of the brain,[84] hence anger is one of the famous 'five stages of grief' model, first postulated by psychiatrist Elizabeth Kübler-Ross* in 1969.[85]

However, whatever the reason, the result was that many people communicated expressions of love and sadness to me, but I didn't share or reciprocate those emotions. Instead, I felt anger.

This was alarming, but led to me appreciating a very important point: just because the human brain *can* share the emotions communicated by others, there's no guarantee that it *will*. The neurological process underpinning the processes of empathy and emotional communication are fundamental in the human brain, but there are also many other things going on that get in their way.

One of the most obvious issues is that we have our own emotions, and these can clearly interfere with our empathy. Case in point: my unspecified anger certainly distorted my perception of the compassion and sadness expressed by others.

A 2013 study, led by Professor Tania Singer of the Max Planck Institute, demonstrated exactly this phenomenon.[86] Subjects were presented with stimuli that were either pleasant (seeing puppies

---

* The model that argues there are five sequential stages of grief: Denial, Anger, Fear, Bargaining, Acceptance. As a neuroscientist, the notion that a profound and complex emotional experience like grief would play out in the exact same way, for everyone, seems far-fetched. Admittedly, Dr Kübler-Ross never originally said the five stages would happen for everyone, or in the same order. Nonetheless, over time that's ended up being the common understanding of it.

while touching soft fluffy things) or unpleasant (seeing maggots and rot while touching gross slime). They then had to evaluate either their own, or someone else's, emotional reaction. If both subjects were presented with the same thing (e.g. something gross), they excelled at gauging the other person's emotional state. But if subjects were presented with *different* stimuli (e.g. one got fluffy puppies, the other gross maggots), they were very bad at accurately gauging the other person's emotional state. They struggled to empathise.

Remember, our brains are demanding yet frugal when it comes to energy and resources. So, if your limbic system processes are already engaged in producing a certain emotion, it's going to take energy and effort to change it to something else, a necessary step in empathy for someone in a different emotional state. If they're in a similar emotional state, it's a lot easier, like how crossing the street is a lot easier than driving across town.

One outcome of this is that our empathy can end up having an egocentric bias. We default to thinking, 'I feel like this, so they must feel like this too', because our own emotions are influencing the process. Luckily, as we know, our brains can also differentiate between someone else's emotions and our own, so can factor this into the process of empathy, like a golfer adjusting the direction of their swing to factor in the wind.

According to this 2013 study (and others[87]), this ability comes from the right supramarginal gyrus,* another important brain region, with many overlaps with the networks and areas responsible for empathy.[88] Correspondingly, a subject's ability to empathise with those experiencing different emotions is drastically reduced if their right supramarginal gyrus is compromised,[89]

* Again, this means we're talking specifically about the supramarginal gyrus in the right hemisphere of the brain. The one in the left hemisphere seems to be more involved in word recognition and similar processes.

be it via damage, too little time to make adjustments, or our own emotional state being overwhelming. Even so, it's hard to deny that our own emotions can disrupt our empathy when we've had to evolve specific neurological mechanisms to counteract this.

Why the other person is feeling the emotions they're displaying is also an important factor. If you perceive that someone's happy, empathy suggests that you'd share their happiness, right? But what if they're happy because they're dating an ex-partner you're still in love with? Or because they've been rewarded for the success of a big work project, even though you're the one who worked round the clock to get it done? In both cases, the other person, from their perspective, has valid reasons to be happy. But you have valid reasons to be very sad, or angry.

Ultimately, there can be many circumstances where, rather than sharing them, you respond to another person's emotions with your own different emotions. These are complementary emotions. They're not the same as the emotion someone else is displaying, but a reaction *to* that emotion.* When you *do* share someone's emotional state, that's a reciprocal emotion.

The issue at the heart of this distinction is that the brain has to take into account a huge variety of extremely complex factors in processing our emotions, and those of others. Sensory cues (body language, tone, facial expression, etc.) can help reveal our emotional states, but there are also the external and associated details of what's going on around us, our knowledge and memory of the situation, who we're with, what they represent, and so on. It's a lot to take in, so, predictably, there are several neurological regions involved in all this, many of which influence our eventual emotional response.

One is that dependable emotional hub, the amygdala. Among its

* We can also experience complementary emotions to *our own* emotions, like how if we feel happiness about something that we know is bad, we often feel shame or guilt as a result. Emotions regularly lead to other emotions.

(many) functions, it enables our brain to determine the emotional aspect of the current social scenario, quickly and effectively, which helps determine which emotion we experience when engaging with others.[90]

For example, if a stranger asks, 'Is that your car?', are they admiring it, and by association, you? Or are they angry, because said car is currently on top of their dog? Our brains recognise this via the other person's tone and demeanour, but our amygdala incorporates this information and determines which emotional response is the most relevant, deciding whether we should be pleased, or apologetic and afraid.

This is still just the emotional facet of the situation, though. There's a lot more detail than that to consider. Where are you? What are the circumstances? Who are the people around you? All these factors, and more, can significantly influence our emotions and empathy, because different parts of the brain process them and feed them into our emotional processing.

This relationship between cognition and emotion is a complex and elastic one, which shifts and adjusts depending on the circumstance and available information. And this arrangement is not infallible. Have you ever tried to cheer up a sad friend, with a joke or cheerful comment, only for it to fall flat, making matters much worse? Or maybe you've mistakenly believed someone was making romantic approaches to you, leading to much embarrassment when you respond? Put simply, the conclusions our brains come to about what someone else is feeling and/or thinking can be *wrong*. When this happens, because our responses, emotional and otherwise, are based on these conclusions, they're wrong too.

One of the reasons why even our sophisticated brains don't always land on the right interpretation of such situations is that what someone's feeling and what they're *thinking* aren't always the same thing. As we've seen, empathy enables us to share in others' emotions, but what is required to work out what someone else is thinking – why they're doing what they're doing – involves a different process: mentalising.*

While there's obviously a lot of overlap, empathy and mentalising are supported by different neurological systems.[91] Mentalising utilises some of the same systems as those that we have seen support empathy (e.g. the superior temporal cortex[92]), but it depends more on brain areas like the medial prefrontal cortex, temporal pole, and ventromedial prefrontal cortex.[93]

Mentalising and empathy can also work against each other. For instance, if we're convinced that someone has malicious intentions (via mentalising), we wouldn't empathise with them if they were sad, hurt, or happy. Conversely, if we're very emotionally invested in someone, we tend to show a great deal of empathy for them,[94] which plays havoc with our ability to rationally work out what they're thinking. It's a depressingly familiar experience, to see a friend manipulated or exploited by a romantic partner, because they're unable (or unwilling) to conceive that someone they love could have nefarious intentions or motives.[95] So, even though they often cooperate, empathy and mentalising can easily get in each other's way.

In a nutshell, our brains are good at *recognising* that other people have their own distinct emotions, but working out *what* these emotions are, and incorporating this information into our reactions and behaviours, is a bigger and more demanding task. We don't have direct access to other people's emotions and thoughts, so must work them out indirectly, via assessment of the information gleaned via our observations. But we *do* have direct access to our own emotional experiences and memories, which

* Also known as 'theory of mind' or 'perspective taking'. Whatever label you use, it essentially boils down to our ability to metaphorically put ourselves in someone else's cognitive shoes.

often means they significantly shape and influence our empathy for others. But because people vary so much, this can mean we don't empathise correctly. Given how empathy underpins a great deal of interpersonal interaction, this can be a problem. If you're not empathising with someone correctly, it'll hamper your ability to connect and communicate.

A striking example of this is the 'double empathy' problem, which can happen when an autistic person interacts with a neurotypical individual.[96] Here, the differing workings of autistic and neurotypical brains can mean that the experiences each person has, and the resulting memories obtained via them, aren't particularly helpful for interpreting the emotional cues produced by the other. It's as if each is displaying their emotions in a language that the other person is still learning. Nobody is doing anything wrong, but both brains involved in the interaction are perhaps too heavily reliant on personal, subjective experiences to make sense of what the other one is doing. Hence, the double empathy problem.

And this leads to one final, somewhat unsettling issue. Even if we leave aside our own emotions, the wider situation, and our flawed cognitive assessments, there can be a simpler explanation for why we fail to feel empathy for someone. Whether you call it prejudice or fear of the unknown, there are times we simply don't *want* to empathise with someone. Because they're different to us, in a way we don't like.

This is hardly a revelation, sadly. It's a fact of life, demonstrated by 99 per cent of all news stories and their corresponding online comment sections. But while it's easy (and bleakly popular) to write this off as humans being fundamentally flawed and irredeemable, there's a lot more to it than that.

We earlier saw evidence suggesting that human empathy is intrinsically altruistic[97] because we show enduring empathy for complete strangers. But this depends on what type of stranger we're talking about. Are they from a recognisable group, community, or ethnicity? If so, do you have pre-existing feelings about that group? If the stranger looks and sound like you or your family/friends, you'd have more positive associations with them. Conversely, if they're from a different ethnicity or culture, or a member of a group you've had negative dealings with before, this has been shown to lead to feelings of wariness and mistrust, so reducing our ability, and motivation, to empathise with them.

In technical terms, it's far easier to empathise with members of our ingroup (the people or community we identify with, that we feel we belong to), than with members of an outgroup (any distinct and recognisable community of people which isn't your ingroup).[98] For example, if a member of our ingroup is suffering, we'll likely show empathy and feel sadness too. But if we see the same emotions expressed by someone from an outgroup, we may experience something more akin to schadenfreude than empathy.[99]

As unpleasant as they can be, these ingroup versus outgroup tendencies can stem from deep within our brains. One of our better known unconscious biases is the cross-race effect,[100] where we're typically better at recognising and differentiating between faces from our own race, compared to faces from other races. This is often considered a sign of unthinking racism (e.g. people declaring, 'they all look the same to me' about people of different races), but it does seem to be a genuine phenomenon.

Admittedly, most people are raised by a family, and as part of a community, that's largely the same race as them. Therefore, our brains get more practice at discerning between members of our own race than those of others. And faces are integral to recognition and emotional expression, so this can affect our ability to show empathy for individuals from other races.[101]

It's not just a quirk of upbringing, though. Thanks to our evolutionary history, the ancient reflexive systems still knocking about in the lower regions of our brains mean that when we encounter someone from an outgroup, it can trigger a rapid response in our amygdala, activating our threat-detecting fight-or-flight response.[102] So, the instant we perceive someone from outside our 'safe' ingroup, our emotional systems are already generating fear and wariness, which skews our overall emotional reaction towards the negative.

This doesn't excuse racism and prejudice, of course. Our more sophisticated cognitive systems can, and will, control and override such unconscious biases,[103] if we're willing to involve them.

Another important point is that how we define ingroups and outgroups, and whether we're likely to empathise with them, is surprisingly hard to pin down, thanks to how complex the brain can be. It's easy to assume it's based on obvious physical differences like race or gender, but our own development and background and experiences can render such things irrelevant; studies have shown that members of a diverse, multicultural community don't seem to have much difficulty in recognising the emotions of individuals from different races.[104] It doesn't mean there *won't* be outgroups we don't think well of, on an almost instinctive level, but what *defines* those outgroups can emerge more from personal experiences and attitudes than anything more innate.[105] This explains why someone can be totally fine with anyone from a different race, but passionately despise people who support a football team that rivals their own preferred club.

Also, it's not that we *can't* feel empathy for someone from an outgroup. It depends on the situation. Have you discovered something in common with them? Does your own group encourage and reward empathy for the outgroup? Not every outgroup is automatically a rival; they could be friends or allies. There are many reasons for us to empathise with those who are 'different'. Evidence even shows that increased exposure to people from outgroups, making them more familiar, can make us more likely to show empathy for them.[106]

At this point, several things have become clear. We don't automatically empathise with everyone we encounter, because our brain takes many other things into consideration. But also, the things that stop us from showing empathy aren't automatic either. It's all determined by a mishmash of influences at the various levels of our brain's workings, which makes things rather confusing and unpredictable. And this arrangement can change over time, as our experiences alter our empathic abilities. But then, these changes aren't always for the better. I'd certainly found this, given how my own confusing emotional reactions had obscured my ability to feel empathy, to appreciate the emotions communicated by others.

However, this isn't intrinsically a *bad* thing. There are many times when empathy is useful, but also many times when a *lack* of empathy is useful. We've already seen that our brains limit just how much emotion we can experience via others, because there are plenty of instances where that's the best option. Being caught up in an angry mob is rarely a positive outcome. If you're a commanding officer leading soldiers into battle, or a nursery teacher supervising a class of wailing five-year-olds, sharing the fear and distress of your charges will seriously hamper your ability to do what's needed.

Clearly, there's a balance to be struck, between too much and too little empathy. So, I wondered, should I be striving to open myself up more to the feelings of others? Or is my reduced empathy actually *protecting* me from further emotional turmoil?

Considering this, I decided to speak to a doctor. A medical doctor. Because if you look into the literature about improving emotional communication and empathy, managing emotions, emotional experiences, burnout in the workplace, etc., a great deal of it is aimed at, and comes from, the world of medicine.[107]

This makes sense. To do their jobs professionally and properly, doctors and related medical professionals (nurses, physiotherapists, etc.) must maintain an emotional distance from the many unwell people they treat. A lot of medical interventions are still quite risky and unpleasant, and administering such things to individuals you've formed a strong emotional bond with is going to be very challenging. Indeed, there are official guidelines concerning medics not letting their own feelings and views interfere with their work.[108] Excessive emotional strain at work can also be very bad for your wellbeing, so constantly feeling keen empathy for many people experiencing health-related hardships is a sure-fire route to stress, burnout, and mental health problems. It's no wonder that medical training so often ends up, by accident or design, teaching aspiring doctors to regulate, suppress, or deflect the emotional experiences they encounter as part of the job.[109]

On the other hand, it's increasingly recognised that being emotionally distant or aloof is a risky approach too. Patients tend to respond negatively to it, making a medic's job harder.[110] Emotions are a big part of thinking, identity, and wellbeing, so ignoring the emotional needs of patients is often self-defeating. That's why so many hospitals also have chapels,[111] or other religious places and services.

All in all, working in medicine regularly seems to involve walking a tightrope between experiencing too little and too much emotion, between having too much and too little empathy for patients. How do they maintain this balance?

To find out, I spoke to Dr Matt Morgan, experienced intensive care doctor, and author of the excellent and eye-opening book *Critical: Stories From the Front Line of Intensive Care Medicine*.[112] Dr Morgan has written a lot about the issue of emotional interactions and connections with patients, and particularly about the dilemma in medicine about how much emotion to show to patients or their families:

Crying in front of patients can show you care, which can help. But it's also something of a role reversal; you're the one who's supposed to be supporting them, and if you're crying it can seem like you're expecting them to support you. On the other hand, being stone-faced when giving bad news can give the wrong impression too.

Clearly what is going to be an appropriate level of emotional engagement differs depending on individual circumstances, but Dr Morgan also pointed to a more prosaic reason that many medical professionals end up tempering their emotions at work:

Part of it is just the time as well. In the modern medical workplace, we've rarely got enough time to do everything we need to do as it is, which leaves us little to no time to recover and process what we've gone through, if it's been particularly emotionally challenging.

It's a point I've heard made many times, and something that always comes to mind when hearing people who have had negative experiences of the healthcare system complain about doctors being uncaring. Is it that 'doctors don't care', or just that 'doctors don't *have time* to care'? Because if Dr Morgan is indicative, doctors certainly do care about their patient's emotional needs:

I have the same ritual whenever I have to break bad news to anyone. I always write the correct names on my hand, check the room, make sure there's nothing on my clothes or shoes, and so on. I want to make absolutely sure there's no chance of me saying the wrong name, or there's no radio or distracting noise playing in the background. These minor details can become much more significant and upsetting when someone's dealing with the emotional impact of the worst news they'll ever get. It could stay with them forever.

A very astute observation, given what we've seen about how powerful emotions amplify memory formation. Unfortunately, even if doctors *want* to be as emotionally considerate and communicative as possible, the nature of the job means they often can't afford to be:

> When you're a doctor speaking to a patient, you need to be as clear as possible. If a patient has died, you can't tell the family 'He's gone to a better place', or 'He's moved on', because they could think 'Where to? What better place? Another ward? That's good, right?' You can't risk doing that to them.

Some may scoff at this, but remember, powerful emotions cloud our thinking. Family members waiting to hear news about a gravely ill loved one are *not* going to be in an emotionally neutral state. Trust me on this.

It's not always a conscious decision on the doctor's part, either. When you're working in a quiet intensive care ward, filled with severely ill people and their concerned family and friends, it would be incredibly jarring to hear someone laughing and joking, so doctors don't. They adhere to the mood of the environment, maybe via simple emotional contagion. This means they're regularly feeling genuinely sombre and sad. This can take its toll, as Dr Morgan explained:

> You do see a lot of emotional wreckage in medicine. I've known many people in the field who have since died by suicide. And weirdly enough, it's often the people who seem the most outwardly happy.

As we've seen, *suppressing* emotions harms wellbeing, but feeling compelled to express *different* emotions to the ones you're actually feeling is worse again. And when you're working in a field as fraught and challenging as medicine, it's no great leap to assume that such harm to your wellbeing can be so bad as to be life endangering. So, how does Dr Morgan deal with it? And can his methods be of any use to those of us in a different field of work?

'Over time, you become used to it. It's familiar, it becomes part of who you are,' he said. I can relate: a similar thing happened to me when I worked with the cadavers. I wouldn't say that was the healthiest approach, though. I'm still not sure what the lasting impact on me has been. But then he added something else:

I do find I tend to react more emotionally to enjoyable things outside work now. It makes me appreciate life a lot more. And when I've had a hard day, or something deeply emotional has happened during my shift, I can avoid reacting, I stay in control and do my job. It's when I get home, and I'm hugging my daughter or speaking to my mother, *then* I can feel the emotional impact of it, it all comes out then. And it's OK, because it's safe to do so. It's good.

Perhaps that's the answer? Perhaps it's not that I'm unable to process all the emotions that have accumulated since my dad's death, it's more that I just *haven't*. Not yet. Largely because I've been cut off from sharing my emotions with friends and family, and sharing emotions, empathising, with those we're connected to, is a big part of the human emotional experience.

But maybe I need that right now? Maybe being influenced by the emotions of others would hinder my ability to make it through this confusing and grim situation intact? Thankfully, though, this isn't a permanent situation. And as Dr Morgan (and an extensive body of literature about people resolving emotional problems many years after they were acquired[113]) revealed, there's no apparent cut-off point when it comes to dealing with your emotions and the hassle they're causing you.

The pandemic is ongoing at the time of writing, but it won't be around forever. If I need to keep a tighter grip on my emotions for now, maybe that's fine? Whatever works. But this could, and probably should, change, as soon as I get to be back in regular close contact with those I care about, who are willing and able to help me with everything I've gone through.

I don't know when that will be, but I do know one thing.

It'll be emotional.

It could very well lead to a dramatic reshaping of my relationships with others. And that's something to be vigilant about, because, it turns out, our emotional relationships and connections are even more important to us than most realise.

5

Emotional Relationships

Once, after the traditional Sunday roast at his house, my father and I had a general catch-up on what we'd been up to. He'd been spending a lot of time helping care for an elderly relative, in decline from Parkinson's disease. Dad was never one for refraining from speaking his mind, so this resulted in a long, impassioned rant about how he now saw that care workers are obscenely undervalued and underappreciated by wider society, that dedicating yourself to looking after someone 24/7 requires copious effort and sacrifice that is so often ignored or dismissed. But people still do it.

'That's what you should write your next book about,' Dad insisted. And to be honest, I felt he had a point. The way people will make immense sacrifices for certain others, often for little or no objective reward: maybe there was something in that.

I've thought about this discussion with my father a great deal since. Partly because the pandemic later catapulted the role of care workers, and how unappreciated they are, into mainstream discourse. Dad was ahead of the curve there.

But the main reason I keenly remember that conversation with my father is because it was the last interaction we'd ever have, face to face. Less than three months later, he was gone. And I got to experience the darker side of caring about someone.

Because in the weeks after his death, the sheer banality of everyone else's lockdown lives seemed an affront to me. How *dare* people bake sourdough, do Zoom quizzes, and go for walks in the sunshine? My father just died, and they're acting like everything's normal! This huge-hearted larger-than-life individual, beloved by many, had been taken from us way too soon, and everyone's acting like it means nothing to them? It was disgracefully disrespectful.

Only, it really wasn't. Yes, my father dying was like having a pillar kicked out from under the foundations of my world. But the key word there is 'my'. My emotional reaction was so potent because I was very emotionally connected to him. Most people . . . weren't.

Sure, Dad was indeed popular, renowned, and loved by many. Nonetheless, most people weren't saddened by his death. They didn't care about him. Because they didn't know he'd existed in the first place. My being angry and upset at them didn't change this harsh reality.

This highlighted a key factor of emotion and empathy that I'd not considered before. Whether we emotionally respond to or engage with someone often depends on how much we *value* them.[1] It's fine to say that all humans are inherently worthwhile (because they are), but we all have our own special group of friends, our close confidants, our family, our *loved ones*.

No matter how you feel about humans in general, our emotions invariably cause us to form close relationships with select individuals. And these in turn dictate our emotional responses and behaviours.

Basically, if I wanted to understand emotions, I needed to understand why we care about others so much. Just like Dad suggested.

Baby steps: how the parent–child bond shapes our emotions

Figuring out my emotions following the death of my father has been, shall we say, a struggle. In so many ways, this is uncharted territory for me. However, one thing I'll say with confidence is this: it would have been significantly more difficult if I wasn't a father myself. Yes, losing Dad has made my own emotional ignorance very apparent, but I was *more* emotionally ignorant before I had children of my own.

This isn't to say I was emotionally distant or aloof as a younger, child-free man. I got angry, sad, or scared about the usual things. I loved my family, my wife, and happily said so when I felt it was required. But that's the point: I'd express emotions, but only ever on my terms. In precise, controlled ways. The idea of doing otherwise felt like 'surrendering control'. After all, I was Mr Science, the brain man. I couldn't be seen to be beholden to my emotions! What would people think?*

But as should be clear by now, our emotions occur when they damn well like, regardless of whether it's 'convenient'. And regarding how our brains work, experiencing, processing, and displaying emotions are not distinct things. Insisting otherwise is detrimental to good wellbeing.[2] If I still had this mindset when my father died, it would have hit me much harder than it already did.

I've got my kids to thank for changing that. I really can't overstate the emotional impact of having a tiny, fragile human placed into my arms by medical professionals saying, 'Congratulations, this is yours forever'. It was . . . intense. I was grinning, panicking, cooing, fretting, dumbstruck but babbling (at the same time, somehow), nervous, bewildered, and much more besides. The rigid mental control over my emotions that I'd built up over the years immediately crumbled, like a barrier of breadsticks trying to stop a freight train.

Because here's the thing: I understood that I was going to be a parent on an *intellectual* level. I'd bought all the pushchairs and bibs, learned all about school catchment areas, been to prenatal classes, and so on. But holding your firstborn for the first time, that's when it becomes 'real', and what it means for you, and your life, is suddenly extremely tangible. And *that's* when the emotions hit. For me, at least.

---

* My time in the anatomy department may have also played some part in this.

Predictably, the overriding emotion I had upon becoming a parent was happiness. A few weeks after he was born, I was carrying my baby boy around the house, trying to get his wind up, and suddenly realised it was Friday night, and I'd usually be out socialising with friends then. I also remember not caring, being perfectly happy where I was.*

Why is it that we experience such intense positive emotions upon becoming a parent? Babies don't do much. They just lay there, gurgling or crying, ruining your sleep, needing constant feeding, and producing noxious substances that you have to clear up. And this goes on for years.

Also, having a baby around is, objectively, an immense burden: psychologically, physically (for mothers especially), and emotionally. They mean a drastic loss of autonomy and independence, significantly increased demands on your energy and finances, sleep loss, anxiety, and more. This is all stuff our brains usually object to strongly. It stresses us out. This presumably contributes to postnatal depression being so common in mothers,[3] *and* fathers.[4] It's no wonder many are apprehensive and fearful about the prospect of becoming a parent, even actively avoiding it.

Yet, despite all this, the love between parent and baby is probably the most intense and enduring emotional connection that can exist between two humans.† This may not make sense from a purely rational perspective, but then emotions and rationality rarely see eye to eye. My suspicion was that something about babies manipulates or influences our brain's emotional systems, to facilitate the parent–baby

---

* That remains the case to this day.

† As ever, this doesn't necessarily apply to everyone. Brains vary considerably from person to person; there's a great deal going on in each one, so you're going to have many instances where this emotional connection between parent and child just isn't there, or isn't what would usually be expected. It's deeply unfortunate for those involved, but it's an inevitable part of life, sadly.

bond, to amplify the positives and suppress the negatives.

But in reality, it's the other way round. Rather than babies exploiting our brain's usual emotional processes, those emotional processes exist *because* of babies. For instance, we've looked at the complex neurological processes behind phenomena like empathy and emotional contagion. But there's one important chemical aspect I've not mentioned yet: oxytocin.

Oxytocin is a relatively simple peptide molecule, produced in the hypothalamus and secreted into the bloodstream by the pituitary gland, but there are numerous neurological connections to these regions, which facilitate oxytocin's use in the brain.[5] Hence, oxytocin is a neurohormone: it functions as both a neurotransmitter *and* a hormone. It has receptors in multiple tissues and areas throughout the brain and body.

You've maybe heard of oxytocin already. It's quite well known, often labelled the 'cuddle' or 'love' hormone. And for good reason: oxytocin levels are higher in couples during the earlier 'infatuation' stages of their relationship,[6] but its presence also seems integral for enduring long-term romantic attachments.[7] It's released in great amounts during sexual activity; it's involved in both the psychological and physiological aspects of lust, like arousal, erection, orgasm, etc.[8] Oxytocin administration has been shown to make men more attentive to/protective of their female partners,[9] or even find them more attractive.[10]

However, the role of oxytocin isn't limited to intimate, romantic interactions. It's seemingly released by *any* positive interaction with another person,[11] making even looking at the face of someone we care about a rewarding act. Quite literally: oxytocin can, and often does, stimulate activity in the brain's reward pathway, the source of our ability to experience pleasure, which explains why we humans often find the company of others so enjoyable.

It would be wrong to say that oxytocin *creates* positive emotional attachments, though. It's more that the action of oxytocin can enhance, increase, and amplify the emotions we experience with regard to other people, thus making us more invested in them.[12] This usually means more oxytocin is produced when we interact with them, meaning we enjoy their company more, and so a positive reinforcement circle is formed. Added to this, oxytocin enhances the encoding of positive social experiences by our memory system.[13] Overall, it's reasonable to conclude that oxytocin plays a key role in determining who we get, and remain, emotionally attached to.

By way of analogy, in a school science lesson I once had to use wires to connect batteries to small bulbs. If the bulb lit up, you'd successfully constructed a rudimentary but workable electrical circuit. However, this lesson was arguably *too* successful, as some of my more mischievous classmates realised that you didn't have to limit the circuit to just one battery, so started adding more, and more. At one point they had a circuit with five batteries* powering the tiny bulb, which, rather than the expected dim glow, was shining intensely, like a fragment of the sun.

If we say the original circuit represents the human brain's social emotion system, and the output of the bulb is our capacity to emotionally engage with others, then the 'additional' batteries represent oxytocin and its effects. And, just like how an over-powered lightbulb can burn out quicker and be painful to look at or touch, the actions of oxytocin aren't always positive.

The effects of oxytocin vary significantly depending on our emotions, context, and the people around us.[14] For instance, oxytocin has been shown to enhance feelings of schadenfreude and envy,[15] as well as emotions and behaviour which lead to us prioritising those

---

* And these were those old-school heavy-duty batteries, like a plastic-wrapped brick with two springy terminals on the top.

we're familiar with, whilst being more suspicious and defensive towards those we don't know.[16] It's perhaps excessive to say, 'oxytocin makes you more racist', but in certain scenarios it seemingly does something along those lines.

Dr Richard Firth-Godbehere describes oxytocin as part of the fuel in 'a belongingness engine', which can cause real suffering in some circumstances, like the intensely negative emotions I experienced in witnessing people behaving perfectly normally following my father's death.

For better or worse, the human ability to emotionally communicate and bond with others is a crucial facet of how we function, and a major part of how and why we're the planet's dominant species.[17] And oxytocin amplifies and sustains these emotional abilities. How did one simple chemical end up being so important?

Well, firstly, while it may be the primary one and gets the most attention, oxytocin isn't the only chemical within us that influences our emotional interactions and connections. Various other substances are involved, but the obvious one to consider is vasopressin. You could describe vasopressin as a sister chemical to oxytocin, both structurally (they're chemically remarkably similar) and functionally. While vasopressin doesn't get the attention oxytocin gets, when it comes social emotional processes, oxytocin and vasopressin overlap and interact a lot.[18] For instance, vasopressin is seemingly integral for forming long-term monogamous bonds in males.[19]

Secondly, oxytocin didn't just appear out of nowhere. In the evolutionary sense, it's been around in some form for hundreds of millions of years. Analogues of it exist in almost all known species, with a wide range of functions, such as regulating the water balance of cells.[20] However, the specific chemicals we recognise as oxytocin and vasopressin are almost exclusively found in one type of creature: mammals.[21]

What sets mammals apart from other species types? Besides a tendency towards fur and hinged jaws, the key feature of mammals is how we reproduce. Mammals grow offspring within their bodies, where they're nourished by the mother via the placenta. After birth, mammals are nursed with milk, expressed via mammary glands in the mother's body. So, mammals are the only creatures to utilise 'proper' oxytocin, and the only species to give birth to and raise live young.* It's not a huge leap to suspect that these two facts may be connected. And that seems to be the case.

If emotionally pleasant social interactions increase our oxytocin levels, giving birth and breastfeeding seem to open the floodgates, sending oxytocin levels through the roof.[22] Oxytocin's prominent role in the birthing and nursing of babies is one of the first things it was recognised for in humans. Indeed, the word 'oxytocin' is derived from the Greek for 'rapid birth'.[23]

Oxytocin both helps to initiate the process of labour, and is released during it. This positive feedback loop floods a mother's system during childbirth, presumably to offset at least *some* of the associated physical distress and discomfort. In some cases, this effect can go overboard, making the birthing process not just arduous and painful, but bizarrely euphoric.[24] Oxytocin is also triggered by the sensory aspect of breastfeeding, increasing a mother's milk supply and expression.[25]

There's an obvious evolutionary benefit to mammals of a chemical that triggers labour but makes it less severe, and also stimulates expression of milk when an infant attempts to feed. But giving birth is not just a physical and sensory process; there's the mental aspect too, giving rise to the mother–child emotional bond, arguably the most powerful we're capable of forming. Like many

---

* As in, they don't lay eggs or spawn, and they rarely wander off and leave their progeny to fend for themselves, like fish and reptiles tend to.

other mammals, human brains have systems in place to predispose mothers to becoming intensely emotionally attached to their offspring, particularly when they're extremely young and vulnerable. And oxytocin is an integral part of these systems.[26]

And this is a two-way arrangement. While the birth process swamps a mother's body with oxytocin, that's doubly true for the baby. Remember, birth is the very first experience a baby's brain will have to deal with, and it's undoubtedly an upsetting one. To go from a warm dark sac of fluid into a world of cold air, bright light, and strange unfathomable noises, surrounded by giants wielding incomprehensible instruments – how could that *not* be traumatic?

Thankfully, newborns have even more oxytocin flowing through their systems than their mother. This can reduce stress, discomfort, and pain, but its most important role is in making us more emotionally open and sensitised,[27] meaning both mother and baby are maximally primed for emotional bonding. This is particularly important for the baby: the bond with its mother isn't just the first bond it forms, it's practically the first thing it experiences *full stop*.

Oxytocin is involved in maintaining and reinforcing emotional bonds, and it's produced in response to skin contact, hence skin-to-skin contact between mother and baby is usually the first thing that happens following the birth.[28] Skipping it can be a factor in postnatal depression.[29] Its release in breastfeeding helps further strengthen and deepen that emotional bond. And these positive effects aren't limited to mothers, either.

Studies have revealed complex networks in the human brain, in both men and women, that regulate and initiate caregiving behaviour.[30] They take in the limbic and cortex regions, meaning they involve the mechanisms underlying emotion production and regulation, as well as complex thought and planning, reward, reflex, and motivation. All these brain processes, working in concert, make us feel very caring towards whatever sets this system off. And the sensory cues produced by babies are the most reliable and potent trigger for this caregiving instinct.[31]

Basically, it seems our brains did indeed evolve to respond, in strongly emotional ways, to sensing the particular physical traits of human babies (the large heads and eyes, the distinct sounds of their laughs and cries, even their smell), producing an instinctive drive to display caregiving behaviour, to want to look after and engage with the source of this stimulation.[32] And oxytocin is a big part of this.[33]

Basically, our brains are hard-wired for strong emotional reactions to babies. While this applies to many mammals,[34] humans take this to extremes. Our young are born in a far more vulnerable and fragile state than most mammals, and take much longer to fully mature (both believed to be a result of the biological demands of the hefty human brain[35]), so infant humans comparatively require much more caregiving, for longer periods. Inevitably, our brains evolved to facilitate this, by making human parental attachment and caregiving instincts especially powerful. However, useful as it may be, it led to some weird outcomes.

For example, there's one member of my family I haven't mentioned yet: my cat, Pickle. Pickle is, as they say, 'a character'. Granted, everyone with a cat talks about the antics of their own particular furry friend, but even seasoned cat owners have described Pickle as 'a bit much'. For instance, we live round the corner from the local school, meaning Pickle has invaded PE lessons, sports day, a talent contest, a staff meeting, a harvest festival, and on one memorable occasion, the headteacher's car.

We've also had neighbours complain to us about Pickle 'bullying' their pets, even the ones with a husky easily eight times his size. The most common greeting we get in our neighbourhood is, 'Oh, so that's YOUR cat?' Add to that the regular gifts of disembowelled wildlife we often find on our lounge rug first thing in the morning, and you'd be forgiven for wondering what exactly we're getting out of having such a pet.

Luckily, there's a remarkably simple reason we like having cats around. As well as being entertaining and endearing, they're *cute*. Why, though? Why do we humans constantly look at these hairy bundles of disdain and slaughter and think 'cute'?

One dominant theory is that, with their small size but proportionately large heads and big eyes, soft fur, limited cognitive abilities, and often playful nature, cats (and other animals we've collectively decided count as pets) have many qualities that we instinctively attribute to babies.[36] That's basically what cuteness is, at the neurological level: our instinctive reaction to things that seem baby-like.[37] Thus, they trigger the same emotion-infused caregiving reflex, making us go all mushy inside, and want to keep them around and interact with them. Essentially, the emotional reaction the human brain has to babies is so potent that it regularly spills over onto completely different species!

And this leads into another weird phenomenon: have you ever encountered something *so* cute, be it baby, kitten, puppy, whatever, that you have the urge to squeeze it, hard, saying things like, 'I just want to crush it!'? Or pinch it? Or even bite it, saying, 'I just want to eat it up!'? If you haven't experienced this, I'll bet you've been around someone who has. It's so commonplace that it barely registers as weird.

But it *is* weird. Cute things are almost always small, vulnerable, and pose no threat to the typical adult human. So, where does this urge to do physical damage to them come from? Obviously, unless you're Lennie from *Of Mice and Men*, people don't normally follow through on these urges. Kittens invariably remain uncrushed by those who find them extremely cute. But however easily overridden or ignored, the fact that such urges exist at all, and are so common, is objectively very strange.

This phenomenon is known as 'cute aggression'.[38] In truth, this experiencing of seemingly incompatible emotional reactions isn't that unusual. People often react to positive emotional experiences in ways usually reserved for negative experiences. People regularly cry when they're extremely happy, or scream when they're intensely excited (the stereotypical example being teenage girls encountering the latest pop heart-throb). Why wouldn't the same apply when we see something cute?

Evidence suggests that, in these scenarios, the cute thing we're looking at triggers such a potent emotional reaction that our cognitive systems become overwhelmed.[39] Our neurological mechanisms cannot keep up with the sudden flood of affect hitting them. They become confused, so just end up producing a generic 'strong emotional response', which includes the negative, aggressive sorts, which aren't really required in this situation.

However, a closer look reveals that there's more order to this apparent chaos than first appears. And this stems in part from the release of vasopressin, which – among other things – stimulates a defensive, protective reaction, rather than the relaxed, cuddly one we normally associate with oxytocin.

Oxytocin and vasopressin interact and overlap a lot, meaning both get to work when we react to babies, infants, and other things we intrinsically feel a strong emotional connection to.[40] One unusual upshot of this, though, is that it leads to activation of the sympathetic and parasympathetic nervous systems *at the same time.*

Just to recap, the sympathetic nervous system is the part of our peripheral nervous system that controls the classic physical components of our reaction to threats, hazards, and other stressful situations, all part of the fight-or-flight response. The parasympathetic nervous system is the yin to this yang; it comes into play to calm the body down and reduce the activity of the sympathetic system when we're more relaxed and content, sometimes known as the 'rest and digest' state.

Usually, these two things are opposing, mutually exclusive. But once again, there are no hard and fast rules as far as the brain and emotions are concerned. Evidence suggests that when we experience cuteness and our innate caregiving tendency is triggered, oxytocin and vasopressin both get to work, and simultaneously activate these two seemingly incompatible reactions.[41]

Our caregiving instincts compel us to nurture the baby, because they're so young and unaware. Oxytocin makes us emotionally sensitive to the baby's needs, more relaxed, more responsive, more engaged, less anxious, less bothered about all the less pleasant elements, like all the bodily fluids and grating noises.

But because they're so small, fragile, and vulnerable, babies don't just need nurturing and caring for. They also need *protecting*. We're motivated to seek out and deal with potential threats and dangers that could pose a risk to the precious young one. Here's where vasopressin comes in, putting us in a defensive frame of mind. But this means activating the fight-or-flight systems, to prime us for challenges and hazards.

While it's difficult to be both tense and relaxed *at the same time*, never underestimate the brain. Something not making logical sense doesn't mean it can't do it. Hence babies and other cute things trigger a deep fundamental reaction in us, which causes us to experience emotions that are both 'Awww' and 'AAARGGH!' And so, we get the weird cute-aggression response.

While vasopressin has a more prominent role in males (hence they're typically associated with the defending of young ones), it's very present in women too. We've talked about the mother–child bond in terms of caring and nurturing, but an equally important aspect is a mother's willingness to rip the face off anything that so much as thinks about harming her child. It's often said that the most dangerous individual of any species (among mammals, at least) is a mother protecting her young. The action of vasopressin is undoubtedly a big part of why.[42]

It's also worth noting that the genes for oxytocin and vasopressin sensitivity can be readily influenced by upbringing, environment, life experiences, and so on.[43] This means that the effect of these hormones, and the subsequent emotions and behaviours displayed towards babies and cute things, can vary considerably between species, between sexes, and between individuals.

Plenty of people just have no interest in children – even their own, in some cases. A lot of this will be down to their stage of life, their upbringing, and the world around them. But at the fundamental levels, an important factor may be that they simply lack the chemical influences that compel others to have such powerful and complex emotional reactions to babies. It's not a flaw or a deficit on their part. It's just how we work.

The thing is, however it's achieved, the attachment between baby and parents/guardians is regarded as an integral factor in how our brains and minds develop.[44] As infants and children, our emotional connection with our primary caregivers often determines what we experience, and how we feel about it. This directly influences how our brains, our personalities, and our identities, are formed.

But the influence of the parent–baby emotional bond goes even deeper than that. We can, and reliably do, form rewarding emotional bonds with individuals who *aren't* our child or parent. We can have lifelong relationships with good friends, be deeply (even worryingly) emotionally invested in belonging to a specific community, and feel emotionally connected to groups of like-minded strangers, often without realising it's happening. And all of this stems from the original mechanism we mammals evolved to keep us invested in caring for our young.

Evolution does this surprisingly often. It can be a ruthlessly efficient process, and if a species needs a new feature or ability, it's often faster and easier to tweak or modify something that already exists, rather than starting from scratch all over again. So, when forming lasting emotional connections became a useful survival strategy for primitive humans, rather than wait millions of years for random mutations and natural selection to cook up a suite of new brain mechanisms, evolution took the existing processes that make us bond with our offspring, and essentially 'expanded their remit'.

The human race itself is an example of this evolutionary tendency. It's often observed that we share around 96 per cent of our DNA with chimps. Despite this, physically and neurologically, we are very different. But *Homo sapiens* have a striking resemblance to *juvenile* chimps. We're largely hairless like they are, walk upright, have a bigger head-to-body ratio, larger eyes, and so on. This even applies cognitively: we're less aggressive and more inquisitive than adult chimps, can retain more information, etc. Many argue that this quirk of evolution is why we're so smart and successful.[45]

It's apparently not even an especially unlikely evolutionary leap, utilising this oxytocin-powered parent–infant bond to make a species more social in general. It's been observed in other social species too, like many rodents.[46] Still, it's hard to argue that we humans haven't taken it to its extremes. We're technically an ultra-social species;[47] cooperation and interaction with others is at the core of so much of what we do, and much of this stems from emotions, our ability to communicate and share them, and to feel them about those around us.

The more emotionally invested we are in someone, the more we empathise, the better we work together, the more we achieve, and so on. Some have pointed out that without oxytocin, and without the ability to form attachments, the human brain as we know it *could not exist.*[48] At the core of all this is our brain's fundamental, instinctive emotional drive to bond with, care for, and protect our young. And in turn, when we're young ourselves, to bond with our parents and caregivers.[49]

That's why it's typically so emotionally evocative, in the worst possible way, to lose a parent. It's not just the loss of someone you know and care about; it's an emotional bond that's been part of the bedrock of your life since day one, that's been integral to who you are and how you've become the person you are. For better or worse.

I may struggle to comprehend and navigate the emotions I felt in response to my father's death. But at least I can understand now *why* I felt them. The parent–child bond really does seem to be one of the most powerful, and influential, our brains are capable of. It also explains why having my own children essentially sent a shockwave through my own emotions, shattering the protective shell I'd spent years thinking was a good idea.

I'm honestly grateful to my kids for making me more emotionally capable of dealing with my father's passing. But then, given our apparent evolutionary history, I should maybe also thank them for the fact that I have emotions *at all*.

Granted, it'll probably seem odd to many to encounter a man being so open about his emotions. But that's a whole other issue that needs to be explored.

Martians and Venusians: are men and women emotionally different?

I revealed earlier that I didn't cry when my father was admitted to hospital. Here's something I'm much more ashamed of: I didn't cry much at Dad's funeral either. Which was weird, as it was the saddest day of my life. I felt I should be crying. I actively *wanted* to. Nonetheless, I was only able to cry later that evening, at home, alone, after my wife and kids were asleep.

This unsettled me; what did it say about the workings of my brain? How *damaged* was I by all this?

But then, thinking back to that grim day, I wasn't the only one. And the pattern was clear. My sisters, stepmother, aunties: they were openly crying. My uncles and me, i.e. the men? Not so much. There was a blatant gender difference.

I'd thought about this a lot after Dad died, because, despite all I've said, our relationship was an odd one. It wasn't acrimonious; we were just very different. Dad was 'old school'. With me, his only son, he could be sympathetic or considerate, sure, but he didn't like showing vulnerability or weakness. Openly expressing emotions – particularly in front of other men, even if they're your progeny – fell under that umbrella, as it does for countless fathers among older generations.[50] Me? I was a modern man. I didn't subscribe to such outdated guff about how men mustn't be emotional! But, if Dad was more comfortable letting our feelings go unsaid, I was fine to indulge that. We both knew the score.

But when I couldn't cry at his funeral, I had to wonder: had I been kidding myself? Many of my preconceived notions about emotions had been completely upended already, so maybe my rubbishing of the whole 'women are emotional, men are stoic' stereotype was also wrong. Could there really be some fundamental difference between male and female brains which explained why I'd had trouble expressing my emotions, while the women in my family did not?

After all, if you want to investigate meaningful emotional relationships, pretty much all of them are between men, women, or some combination/variation thereof. If men and women go about emotional matters in disparate ways, this will have significant impacts on the emotional relationships we can engage in, and our experience of them. But *are* there significant differences between the emotional workings of the brains of men and women?

To begin with, there undeniably *are* obvious differences between men and women. We're typically different sizes, have differently shaped bodies, different genitalia, different distributions of hair, different lifespans, and so on.

Many of these are surface-level features, though, and thanks to the tremendous variation between individuals, there's often a lot of overlap. Many women are taller than many men. Many men live longer than many women. Some men can't grow facial hair, some women can. With that in mind, it may help to focus on the fundamentals, as in, the cellular and chemical levels.

There are the sex hormones, oestrogen and testosterone, and their numerous associated chemicals. They're a big part of our sex and sexual identity. Quite literally, for us men. At conception, all human foetuses are default female. But if there's a Y chromosome in the DNA of said foetus, at nine weeks it starts producing testosterone,[51] which triggers masculinisation, the development and acquisition of male properties.* Men have testosterone in their bodies from this point on, with levels of it spiking at the onset of adolescence, leading to development of classic male traits, like increased bone and muscle mass, body hair, deepening voice, and so on. Oestrogen has a very similar role in the development of women, triggering development of the classical female physical traits,[52] particularly secondary sex characteristics.†

However, these vital hormones (and associated chemicals), aren't restricted to either sex. Oestrogen is found in men, testosterone is found in women, and both have important functions. But testosterone has a more *prominent* role in male development, and likewise oestrogen for women. It's like we saw earlier, with

---

* Therefore, if you're a man who believes that life begins at conception, you should also accept that, at some point in your existence, you were a woman.

† Meaning, the physical properties we have that evolved to 'aid attracting mates', but which aren't involved directly in the reproductive process, like facial hair and broad shoulders in men, permanent enlarged breasts in women, and so on.

oxytocin and vasopressin: both are active in the brains and bodies of all humans, but men tend to do more with vasopressin, and women with oxytocin.

So, even at the chemical level, men and women aren't *quite* so distinct as one might think. And if there's overlap here, wouldn't we expect the same at the neurological level? That's how it usually works, in my experience, given how our brains are significantly more flexible and variable than our physical bodies, and the chemicals which shape them.

Despite this, the belief that men and women are significantly different, and have different brains, is pervasive, stubborn, and widespread. It would take a whole other book to separate the evidence-based scientific reality from the innumerable and deeply ingrained assumptions about gender differences, in the brain and beyond, found in our culture.

Luckily for me, someone's already written that book. That's why I spoke to Professor Gina Rippon of Aston University, expert in cognitive neuroimaging, and author of *The Gendered Brain: The New Neuroscience that Shatters the Myth of the Female Brain*.[53] It provides an extremely useful and eye-opening look at the familiar, widely held beliefs about how distinct the workings of men and women's brains are, and just how drastically these beliefs differ from what the actual scientific evidence suggests.

Professor Rippon is, understandably, acutely aware of just how deeply entrenched these views are in our society. They're found in practically every sitcom and commercial, countless films and books, endless comedy routines, and more. Unfortunately, they aren't just limited to the world of fiction, as Professor Rippon pointed out:

Every time there's a study about sex differences in the brain and it gets picked up by the mainstream press, the way it's covered is very telling. It's always headlines like 'At last, the real difference between male and female brains uncovered' or 'Scientists confirm that men and women have different brains'.

This sort of phrasing in news reports reveals the underlying assumption; that male and female brains *definitely are* different, and it's just a matter of pinning down these differences. But the thing is, there's shockingly little scientific evidence to support this. Certainly not enough to justify it being so familiar and widespread a belief. Unfortunately, it means any study which even hints at tangible sex differences in the brain has a much greater chance of getting media coverage, much more than those which say otherwise, or provide a more nuanced picture.

It's hard to argue that countless people do indeed believe that men and women have different brains, despite what the science says. However, looking back over our history, it would be fairer to say this belief doesn't exist *despite* science, but *because* of it. To an extent.

Due to various cultural factors (in the Western world, at least), for a long time, science was dominated, and thus shaped, by privileged white men.[54] Whatever you think of such individuals, they're not normally associated with progressive, egalitarian thinking. In any case, having *any* field dominated by people with a very limited range of traits and characteristics is invariably a bad idea. It leads to things like groupthink, where members of a community end up thinking or believing things which aren't logical, rational, or evidence based, because they conform to the attitudes and beliefs of the group. It's a well-established phenomenon.[55]

This explains why, for a long time, and despite no robust evidence, the (wealthy, white, male) scientific community genuinely believed men and women had different brains. Indeed, the history of science itself is, ironically, often cited as proof of the inherent difference between men and women. After all, if most famous, world-changing scientists were men,[56] men must be innately better at the qualities science requires, like analysis and reasoning, right? This suggests they have brains better suited for such tasks, whereas women don't. It makes logical sense.

Except it doesn't. It only makes sense by completely ignoring the wider context. Saying, 'Historically, most scientists were men, so men must be inherently better at science than women' is like saying, 'Money is how we reward success, and the children of billionaires typically have the most money. Therefore, they must be smarter and harder working than everyone else, so we should all defer to them and let them run our governments.'* This conclusion overlooks countless factors and variables that result in this outcome.

In any case, for a long time, the scientific community, despite the absence of evidence, 'confirmed' that men and women have fundamentally different brains. Given the role and perception of science in wider society, it's no wonder the belief is so common, so entrenched.

The real problems with this stem from how it's not just the belief that men and women's brains are *different*, but that men's brains are *superior*. The assumption that women are physically and mentally inferior is infused throughout much of history.[57] Women weren't deemed smart enough to vote;[58] it was believed that reading books would render women infertile;[59] my own wife went to a school which was one of the first in the UK to teach women maths, as it was previously assumed that this would cause their brains to 'overheat'. And so on.

These beliefs may seem ridiculous now, but female inferiority being an accepted part of science led to many terrible outcomes. Take

---

* I'm aware that this genuinely happens often in the modern world. This is both ludicrous and depressing.

hysteria, a term currently used colloquially to describe someone being overly emotional and irrational. It was once an official diagnosis. It's derived from *hystera*, the Greek word for womb. Ancient Greeks believed poor health and mental disruptions in young women were caused by the womb becoming detached and wandering around the body, throwing things out of whack.

This (preposterous) conceit persisted in the scientific establishments throughout Europe for *centuries*.[60] And logically, because it's caused by a wandering womb, it was believed that *only* women could become hysterical. So, when men displayed 'hysterical' traits (a regular occurrence), it was never regarded as such, or was instead attributed to things like 'flabby wasted testicles',[61] which are apparently the same thing as a womb.

It gets worse. Another dark chapter in the history of science was the lobotomy, a surgical procedure which severed connections between the frontal lobe and the rest of the brain, ostensibly to alleviate the disruptive symptoms and issues of severe psychosis, or similar conditions.

In fairness, even during their heyday, lobotomies were always controversial. Sure, they typically did indeed reduce the disruptive aspects of psychosis, but usually by leaving patients in a worse state overall, given how they'd *literally had their brains sliced up*. It's like if you take your car to a mechanic because the engine is making a grinding noise, and he just rips the engine out. Granted, the car is much quieter now, but you'd be hard pressed to say he'd 'fixed' anything.

Even so, many prominent scientists still endorsed and championed the procedure. In the past, people have been very alarmed when I've told them about this, because you'd think that anyone who claims that forcing a spike into the base of a patient's brain, via the eye socket, and wiggling it around (which is exactly how many lobotomies worked) was a viable medical procedure would be met with scorn, ridicule, or alarm from the scientific establishment. Someone actually *doing* it, repeatedly, would presumably be arrested, not given a Nobel prize. Nonetheless, the latter genuinely happened.[62]

Here's why this is relevant: during the era of lobotomies, most psychiatric patients were male. But *significantly* more lobotomies were performed on women.[63] There's no logical justification for this, unless you incorporate unchallenged assumptions that women and their brains are inferior, and therefore more 'expendable'. You're losing less by lobotomising them, because women are lesser, and can be treated (and discarded) as such.

Again, it wasn't superstitious cavemen doing this, stabbing women in the head to drive out demons after seeing a strange cloud in the moonlight. It was qualified, respected, and influential scientists. But ones from a culture that believed women were lesser. Accordingly, they shared this belief, and, via their works and influence, validated and sustained it.

The history of mental health and psychiatry is sadly riddled with stuff like this,[64] and while the modern science world is far more rational and evidence-based than previous decades, there are still many instances of women in the field experiencing bias, prejudice, and dismissal, often from male colleagues and peers acting on deep-rooted prejudices, rather than logic or evidence.

It would be one thing if these damaging views about the inferiority of women were restricted to the interactions of the scientific community and didn't spill over into the science practice itself. Unfortunately, that's not the case. An example of this that Professor Rippon highlighted for me was the influential theory that autism is a consequence of an 'extreme male brain', introduced and promoted largely by the work of Professor Simon Baron-Cohen.[65] As she explained it:

This theory obviously assumes that there is such a thing as a male brain and that it is hard-wired for certain aspects of behaviour which are both stereotypically male and more likely to characterise autism.

More specifically, the theory argues that people's brains are better at either systemising,* or empathising, and those with autism are often much better at systemising than empathising.[66] And because male brains, and therefore men, are better at systemising, while female brains are better at empathising, an autistic brain can be described as an 'extreme male brain'.

While this may sound logical, many have serious concerns about this whole rationale, and how it came about. Some note that the underlying research this theory is based on is flawed, or inadequate.[67] Others say that the very premise itself is highly dubious, as both men and women are extremely variable in how adept they are, or aren't, at systemising and empathising.[68] If one sex really were 'hard-wired' for one or the other, you'd see far more consistency between individuals. As a result, many strongly object to the labelling of classic autistic traits as 'male'. As Dr Rosalind Ridley put it in 2019:[69]

The attribution of an 'extreme male brain' to a female with autistic symptoms is philosophically comparable to describing a very tall women as 'having extreme male height' because men are, on average, somewhat taller than women.

The point here is that, even if those with autism do typically show traits more commonly found in men, that doesn't mean

* Analysing, deducing, recognising, or constructing patterns. Creating 'systems' of thought, essentially.

they're specifically 'male' traits. Women tend to live longer than men, but a long-lived man is never said to have a 'female lifespan'.

Despite all this, extreme male brain theory remains very influential in the world of autism understanding. This has led to unhelpful outcomes, like an ingrained assumption that women are seldom autistic, which means women with autism are often overlooked, misdiagnosed, or flat out ignored. Many women I know are all too familiar with this matter, having received an autism diagnosis surprisingly late in life, and efforts at combatting this have been going on for over a decade.[70] Arguably, none of this would be necessary, if not for the extreme male brain theory.

I eventually realised that there was a common theme to many of the deeply unfair beliefs and attitudes women have been subjected to for centuries. The supposed 'unsuitability' of women for science and other intellectual pursuits,* the extreme male brain theory of autism, the stuff about hysteria: it all points to the same underlying assumption, i.e. that women are fundamentally more *emotional* than men.

The thing is, although it's often proven to be ridiculous, illogical, and damaging, this assumption keeps resurfacing. Could it be that there's some underlying truth to it, and it's more the way it's *applied* that's harmful? Nuclear weapons are terrifying things, but it doesn't mean nuclear physics is wrong. Is the same true of men and women having different ways of dealing with emotion? My experiences at my father's funeral suggested there's some sort of emotional gender divide, and many people would support this idea. So, *is* there a scientifically valid difference between how men and women process emotions?

If there is, it would suggest there are tangible differences in the emotional systems of men and women's brains, ones we can,

---

* That the originators of things like science fiction and computer programming were predominately women seldom comes up when this claim is being made. How odd.

hopefully, detect and observe. However, efforts to pin down these differences (assuming there are any) have proven difficult, for various reasons.

For example, many studies have put men and women in scanners looking for any notable distinctions between their respective brains. And quite often, they find them. So, there we go. Case closed, right?

Not quite. Because men are typically bigger than women, meaning male brains tend to be bigger on average.[71] So, if you were to compare, say, a man's amygdala to a woman's, the man's will typically be bigger. Based on what we know about the amygdala, wouldn't this suggest that men are *more* emotional than women, not less?

Again, not exactly. Ample data shows that any detectable impact of brain size on mental abilities is minimal at best.[72] More contemporary studies reveal that many (albeit not all) results suggesting differences in the structure of male and female brains can be explained by men having bigger brains.[73] So, a straight like-for-like comparison of the size and layout of male and female brains, and their components, can't really tell us a great deal.

This is particularly true when we're talking about emotional abilities. We've seen that emotions are produced, modified, influenced by, and trigger responses in, multiple parts of the human brain. Trying to pin down the specific neurological regions responsible for emotion is like trying to locate the exact centre of fog by wandering around inside it. This makes meaningful comparisons between the emotional abilities of male and female brains even trickier.

This hasn't prevented people from trying, however, and some studies have borne interesting fruit, often by focussing on specific emotional abilities and properties, particularly empathy. Several studies suggest that what goes on in male and female brains, in situations where empathy or other emotional abilities are required, can markedly differ.

One study reported that, when presented with infants crying or laughing, women showed reduced activity in the anterior cingulate cortex, while men didn't.[74] This is an area with many important emotional roles, including recognising, sharing, and consciously reacting to, emotions.[75] In this scenario, deactivation of the anterior cingulate cortex could mean you're far more inclined to comfort the child, or otherwise prioritise their needs over your own, because your own emotions aren't being prioritised by the brain region usually responsible. Does this difference in male and female responses explain and confirm the oft-cited 'maternal instinct' that all women supposedly have? Not quite. But it arguably leans in that direction.

Another study found that men and women use different parts of their brain to regulate emotions. Men who were better at regulating emotions showed more grey matter (the neural tissue largely responsible for processing, for making stuff happen) in the dorsolateral prefrontal cortex, while women who were better at emotional regulation had relatively more grey matter in a suite of regions extending from the left brainstem to the left hippocampus, the left amygdala, and the insular cortex.[76]

There are many interpretations of this, but one implication is that men, with more grey matter in complex cognitive regions, use conscious, *deliberate* mechanisms to control their emotions, while women, with more grey matter in the subconscious limbic areas, do it more instinctively. 'At the source', so to speak. This would suggest that women are fundamentally better at producing emotions, while men are better at controlling or suppressing them.

Similar research showed that women display greater amygdala activity in response to negative emotional stimuli, while men show the opposite: greater amygdala activity in response to positive emotional stimuli.[77] This could mean women are more sensitive to negative emotions, and more affected by them. Alternatively, given how amygdala activity is most often associated with fear and danger, it could mean men see positive emotions as a *threat*. Perhaps that explains why men (like myself) tend to be so 'closed off', to equate expressing our emotions with being vulnerable.

And what about our old friends, oestrogen and testosterone? There are many structures in the brain implicated in the regulation and processing of emotions, and one quality most of them share is that they're particularly sensitive and responsive to oestrogen. This means oestrogen strongly influences the activity in the emotion-processing parts of the brain.[78] Oestrogen also stimulates and enhances the activity of oxytocin, and we've seen how important that is in establishing emotional bonds and connections.[79]

Meanwhile, evidence suggests testosterone reduces activity in connections between the rational, self-controlling processes in the prefrontal cortex, and the more fundamental emotional activity in the amygdala.[80] This may be behind the cliché about men getting more aggressive, and acting less rationally, when challenged. Their testosterone is up, and this reduces their emotional self-control.

Linked to this is the fact that testosterone stimulates the expression and action of vasopressin, oxytocin's sister molecule, that induces more defensive, protective behaviour, rather than emotional openness and engagement. This also supports the common contention that women are more emotionally expressive, while men are closed off.

Taking all this together, it may seem clear that men and women *do* indeed have different brains, different emotional mechanisms and abilities. And many people, including top scientists, agree with this conclusion.

Not me, though, because if you look closer, the reality is by no means as clear-cut as one might hope. For the sake of the narrative, the studies I just discussed were the result of deliberate cherry picking on my part, to find some of the more tantalising results that suggest tangible differences. However, for every study which says there *is* a difference between the emotional properties of male and female brains, there's one concluding that there *isn't*, so the overall picture remains murky.

To try to address this, a 2017 study looked at multiple relevant experiments and assessed the pooled data from all of them, to see if any clearer trends emerged.[81] Their results suggested that, while some studies do indeed show tangible differences between men and women with emotional tasks and stimulation, this could usually be attributed to the way the experiment was conducted. Specifically, the experiments which showed a clearer difference between men and women? They were ones where the participants *knew* they were taking part in a study about emotions.

This *matters*, because we're all raised in environments and cultures where it's assumed, and expected, that men and women have different emotional tendencies and abilities. And because of our highly social nature and adaptable brains, we'll often conform to these expectations when we think someone's watching, without even realising.[82] This leads to the ironic situation where the *belief* that men and women are emotionally different is so pervasive that it's actively interfering with efforts to study it properly! In a sense, this even applies at the *chemical* level.

Take testosterone. We all know what testosterone does to us guys, right? The more testosterone you have, the more manly you are. It makes us aggressive, confident, competitive, violent even. Because we males evolved to fight and dominate others, and it's only the restraining nature of our modern world and the expectations of society that keeps these instincts in check. Testosterone ramps them up, makes us men more like what we really are underneath. Or . . . does it?

A 2016 study had male subjects play a game that involved

punishing or rewarding others for their actions. Some subjects were injected with testosterone beforehand. Conventional wisdom would assume that those who had received testosterone would punish others more and reward them less. However, much of the time, they actually treated competitors *better*, more fairly, than those who didn't receive testosterone.[83]

Modern research suggests that this is because the principal effect of testosterone is not, in fact, to make men more aggressive and 'macho', but to make us more aware and protective of our status.[84] Our brains are constantly comparing ourselves to others and calculating where we stand in the social hierarchy.[85] Testosterone amplifies awareness of our social standing and motivates us to protect or advance it.

If we were a species like chimps, where males establish status and dominance by constantly beating each other up, then yes, testosterone making us more aware and sensitive to our status would correspondingly make us more aggressive and violent. We're not chimps, though. We're humans. We're cognitively complex, and *ultrasocial*. When it comes to achieving social status, we've a far wider range of choices than straightforward violence and aggression (although that's still an option). Our brains recognise and value things like intelligence, cooperation, affability, competence, and more.[86] These things shaped our evolution, after all.

Basically, we *like* it, on an instinctive level, when others behave considerately. It's emotionally rewarding, so those who do so are well regarded. So, testosterone boosts our evolved tendencies towards valuing fairness and justice[87] and can make us more considerate and respectful. Because such prosocial behaviour reinforces, or increases, our social status.

A related study into this phenomenon produced remarkably interesting results. Again, this study assessed how subjects treated others when told they'd been administered testosterone, compared to those who weren't.[88] Predictably, those who'd been told they'd been given testosterone behaved far more unfairly, becoming aggressive and overly punitive to others, whenever the opportunity arose.

Here's the kicker, though: although many of these subjects were told they'd been administered testosterone, they *hadn't* been. The only thing driving their overly aggressive behaviour was their belief that testosterone makes you more aggressive. Meanwhile, subjects who were given testosterone *without* being told behaved more fairly and considerately towards others. It seems that when the assumptions about testosterone, masculinity, etc. aren't involved, testosterone makes us *nicer*.

And as a further plot twist, all the subjects in this study were *women*. (Other studies showed the same effect in men, too.)[89] This is profoundly important. Because it demonstrates that while the sex hormones are present in different quantities in men and women, and can have different degrees of influence, they can affect the emotions and emotional behaviour of both in essentially the same way. This strongly implies that the brains of both sexes are more *similar* than different, because testosterone couldn't affect women like this if female brains didn't have the necessary neurological regions and receptors in place to recognise and respond to it. Ditto oestrogen for men.

This isn't to say there are zero differences between men and women's brains. Numerous studies have uncovered several, as we've seen. But even here, it's not so clear-cut, because you have to ask *why* are there these differences? The human brain is a very flexible, adaptable organ, and a mature adult brain will have been shaped by decades of life experiences, a vast chunk of which will have involved emotions and sex or gender roles. Or both. How can we be sure that these are fundamental differences, evolved over millions of years, and not our flexible brains adapting to a lifetime of experiences where these differences are widely assumed to exist, and therefore imposed? Is it nature, or nurture?

Take the data suggesting that men consciously control their emotions more than women. Does this happen because male brains are better set up to do that? Or is it that modern men are constantly told, directly or inadvertently, to control their emotions, so over time *their brains adapt to do this*, and that's what we're seeing when we scan their brains?

It's like the classic experiments which showed that the hippocampus, a brain region essential for spatial navigation, is significantly larger than average in experienced London taxi drivers.[90] These studies demonstrated that the brain, like a muscle, changes shape and structure in response to how much or how little it's used. But nobody has ever made the argument that the taxi drivers originally became taxi drivers *because* their brains had oversized hippocampi. That would be a mind-bogglingly unlikely coincidence, on a par with winning the lottery eight weeks in a row, but never getting your money because each time your ticket gets struck by lightning.

I was once asked, if I could do literally any experiment I wanted, with no restrictions or limits regarding funding, resources, technology, or ethics, what experiment I would want to do. I've thought a lot about that since, and eventually concluded that I'd like to resolve the 'male vs female brain' issue, hopefully once and for all. And here's how I'd do it.

Create human foetuses in a lab. Let's say a thousand, to provide decent statistical power. Make sure half are female, half are male, or XX and XY,* because they're just foetuses at this point. You'd then use advanced incubators to grow them into full-fledged humans. Each one would go through the exact same process, given the same chemicals and nutrients at the exact same time, and so on.

---

* I don't know how you'd do this, but then it's a thought experiment with no technical limitations, so it doesn't matter.

Once they're sufficiently mature, hook each baby up to a *Matrix*-like simulation, where they experience a virtual reality environment, indistinguishable from the real thing. Then they all live the *exact same* life. Same home, same parents, same culture, same life events at the same times, same people around them, behaving the exact same way in every situation, or as near to that as is feasibly possible.

Then, after 25–30 years, I'd do a detailed structural scan of every subject's brain, and compare the male brains to the female ones. Because each brain has experienced the same life, they should have developed and been shaped in the exact same ways. So, if there are still consistent and significant differences between the male and female brains, these are far more likely to be innate, fundamental, nature rather than nurture.

Of course, such an experiment is technologically impossible, and morally abhorrent even if it weren't, so I wouldn't actually do it, even if it were an option. So, for now, we're stuck with this nebulous uncertainty about how male and female brains differ regarding emotions (and everything else). Professor Rippon perfectly summarises the issue here:

It's not that there are absolutely no differences between male and female brains, because there surely are. But structural differences aren't the same as functional ones. The point is, even if *all* the available scientific data that points to there being differences between male and female brains were accurate, it still wouldn't be nearly enough to validate all the widespread assumptions and beliefs about how men and women think and act differently.

That, I feel, is the crux of the matter. Even if there are tangible differences between men and women's brains regarding how they handle emotions, these differences are by no means sufficient to explain, and justify, the different emotional *expectations* placed upon the different sexes. Nonetheless, these expectations are now so embedded in our society that they're essentially self-sustaining. A naïve young brain will develop and adapt in response to what it observes and experiences, and a brain that's experienced a lifetime of reinforcement of the message 'You're a woman, you're too emotional/you're a man, you mustn't show emotions' will be shaped by this, in ways that will show up when you explore the relevant neurological workings.

And while this situation has clearly had countless negative consequences for women, it's not done men any favours either. Cajoling half of the population to control and bottle up their emotions can only have unhealthy results. Emotional suppression in response to negative experiences is a big factor in the risk of suicide,[91] and the evidence is clear that, while more women than men reportedly experience conditions like depression,[92] men are far more likely to die by suicide.[93] Is this down to fundamental differences between how men and women process emotions? Or is it the result of these widespread beliefs and biases about men and women?

Women's lives being shaped by prejudice and countless uphill struggles through no fault of their own logically would result in more cases of depression. Meanwhile, men being actively discouraged from ever expressing their emotions* means that, as well as being less likely to admit to depression and seek help for it for fear of seeming 'vulnerable', they never get the chance to get better at processing and dealing with negative emotions when bad things happen, as reliving and displaying your emotions is a key part of the process. Therefore, men may be less able to deal with the emotional fallout of tragedies and traumatic experiences, which would lead to more making the ultimate fatal step of ending their lives

* Unless it's anger, because anger is 'manly', for some reason.

altogether. This is certainly a feasible mechanism for the differing depression and suicide statistics.

Unfortunately, despite the lack of definitive evidence for them, these beliefs about differing male and female brains don't look to be going away any time soon. They're just too deeply ingrained. Even many modern scientists are still certain that they're essentially correct and are working hard to prove it. Of course, many of these scientists are men who, when faced with challenges to their views that men aren't emotional, can get rather angry or upset.* It's an amusing irony, if nothing else.

On the other hand, one thing I've learned is that I should be less dismissive or flippant about the idea that men and women process emotions differently. Because our lives shape our brains, and if men and women have different experiences regarding what they're allowed to do and what is expected of them, in the emotional sense, their brains will eventually reflect this. It explains the surreal chicken-or-egg scenario my efforts to get to the bottom of this matter ended up at.

So, even if I ultimately know and accept that men, like myself, at the neurological level can and should be just as emotional as women supposedly are, I've still lived a life filled with countless experiences, both subtle and overt, which reinforced the message that this isn't the case. I've internalised the message that I, as a man, should be stoic, and 'tough', and not let my feelings show. As I found when I couldn't cry at my father's funeral until nobody was looking, this sort of programming runs very deep, and overcoming the barriers

---

* Indeed, there's even a (male) US professor who constantly sends angry hectoring emails to anyone who publicly downplays the differences between male and female brains. I've had a couple of messages from him myself, and will likely get more after this is published. You can describe this behaviour in many ways: passionate, dedicated, committed to a cause. But is it perfectly rational? No. There's clearly a lot of emotion involved.

it installed in my brain was a lot more work than I'd have thought.

As far as I can see, it's not that my emotions are stunted because I'm a man. Rather, because I'm a man, my emotions were stunted *by society*. And when an issue is society-wide, it takes more than one person to do something about it. Many, many more.

I'm doing what I can, though. Maybe by getting this out there, it'll help people, particularly men, become more emotionally open and aware, in ways that are healthier overall? I can but hope. I've still got work to do, though. Writing this down in print is actually the first time I've ever told *anyone* about these experiences. I just haven't been able to bring myself to share them openly, face to face, with another person yet. I'm guessing many members of my family may read this and be shocked and alarmed by it, and want to talk to me about it.

If they do, at least I'll be ready now.

For better or worse: how romantic bonds form, change, and break

Even in the darkest times, it seems to be human nature to look for any positives.* And I'm no different. Ergo, I ended up telling myself that, as brutal as this whole thing has been, it's taught me a lot about myself and my emotions, which should make things easier if something like this happens again. Although, another dark positive to take from this is that losing your father is, by definition, a once-in-a-lifetime event – nothing this bad could happen to me again, right?

Actually, not right. I'm never one to make claims without checking the evidence first, and according to the science, worse fates could still befall me. In 1967, psychiatrists Thomas Holmes

---

* Or, in some cases, for people to tell those suffering to look for positives – to 'take comfort' in good memories, etc. Personally, this feels like they're saying, 'Your grief is awkward and distressing and I don't know how to fix it, so could you just stop?' I get where they're coming from, but it isn't exactly helpful.

and Richard Rahe explored 5000 patients' medical records, to see if there's a link between stressful life events and development of illness. They found that there is, and this discovery led to the development of a list of forty-three common experiences, ranked from most to least stressful. This list is commonly known these days as the Holmes and Rahe stress scale.[94]

'Death of a close family member' is number five on the scale, scoring sixty-three out of a possible one hundred. Fourth is imprisonment, also scoring sixty-three. But in third, second, and first place, it's marital separation, divorce, and death of a spouse, scoring sixty-five, seventy-three, and the maximum one hundred points, respectively.*

The implication is clear: all things being equal, losing your romantic partner is the worst thing that can happen to you. And that's the science saying this, not some soppy ballad. Marital separation means losing your partner is a distinct possibility. Divorce makes it official. Death of a spouse is the ultimate expression, as the person you love is not just gone from your relationship, but from the world entirely.

I'll confess that this surprised me. A romantic partner dying would be undeniably horrific, I'm not disputing that, but why's it *so much worse* than loss of a parent? Your parent raised you, is an integral part of your world your whole life, is the person who made you who you are, and you seldom get to seek out new ones.

By contrast, breakdown of even long-term romantic relationships is a common occurrence, and many modern humans will have multiple enduring romantic relationships in their lives. With that in mind, why would divorce and separation be *more*

---

* It's not an exhaustive list; it only features things your average person is likely to experience. Things like war, major accidents and injuries, natural disasters, etc. are statistically unlikely to appear in the medical records of those living in a developed First World nation, which the scale is derived from.

traumatic than losing a parent?* What gives?

It's self-evident that being in love with someone is significantly emotionally rewarding. A loving relationship is a good predictor of general happiness and life satisfaction.[95] Even just being married is linked to generally better wellbeing.† [96]

And when I thought about it, I realised that, as painful as losing my father was, I still 'got through it'. It cost me, but I'm still here, still going. On the other hand, if I'd lost my wife, I'm 100 per cent sure that would have destroyed me, and I wouldn't be writing this now. She's the most important adult in my world, and the idea of being without her simply refuses to stick in my brain. Maybe because, at this point, I've been her partner/husband for the majority of my life, so it's genuinely hard to imagine being anything else.

But then, I was my father's son for my *entire* life. Does this mean I love my wife more, or that I didn't love my dad that much after all? Or did my wife turning up somehow usurp my love for my parents?

No. Because that's not how it works. Love isn't a simple finite resource sitting in your brain, like a stash of coins that people can win, and whoever currently has the most goes to the top of your emotional scoreboard. Love is much more complex than that.

To begin with, what is love? Presumably, most people would say it's an emotion, hence I'm talking about it in this book. However, while most relevant scientists would agree that love is an emotion, it also has many features that the more common or 'straight-forward' emotions don't.

---

* This applies specifically to adults. There's a version of the Holmes and Rahe stress scale for children and teens, and death of a parent is top of that.

† Many relevant studies specifically refer to marriage, but no doubt the same emotional and psychological effects of love will happen in long-term unmarried relationships. They don't depend on the union being recognised in law.

For instance, we can become angry, afraid, happy, or sad, immediately after whatever happened to trigger that emotion. But, despite how many fictional portrayals insist otherwise, instantly falling in love at first sight is extremely rare, if it happens at all. Sure, you can immediately find someone beautiful, and be physically attracted to them. But actual love is a potent and demanding process for the brain. A brain that fell in love at the drop of a hat would be like a complex security system putting a building in total lockdown every time a fly bumps into a window. To genuinely love someone, you typically need to know enough about them, be aware of their qualities, and find them very appealing. It's difficult to achieve this when you've only looked at them once, briefly.

Love is also more focussed than other emotions. A specific incident may make you angry, but the anger endures beyond it, and can be directed at completely unrelated things, like inanimate objects that just happen to stop working while you're already furious. Similarly, good news can make us happy, and this happiness often suffuses our thoughts and actions for the rest of the day. But while people experiencing love may be quite giddy and euphoric in general, they seldom start falling in love with random things, like puddles and chewing gum and traffic wardens. Their love is typically directed at a specific person. And *only* them.

Love also seems to be more enduring than other emotions. We can be angry or sad or afraid for a spell, but we return to a more neutral emotional state relatively quickly. But if we love someone, that feeling of love can last weeks, months, years, our whole lives.

Clearly, when we experience love, there's a lot going on in our brains, more so than with more typical emotions. Accordingly, many scientists describe love as a 'complex emotion'.[97] It has potent emotional elements, that's undeniable, but there are also many more aspects in the mix. Indeed, so complex is it, that the scientific literature recognises several different types of love.

Most people, when they hear the word 'love', would presumably first think of romantic love, the love between two people* in an intimate relationship. However, we typically say we love our parents, but we don't say we're *in love* with our parents. That would be weird. Because it's not that type of love. Indeed, the language we use about love itself reveals that most people acknowledge that 'romantic' isn't the only kind.

There's companionate love,[98] which is the love that exists between friends: individuals with whom you have a positive and affectionate relationship. You enjoy their company, care about them, value their insight and wellbeing, but there need not be any romantic element to it, any attraction or desire for physical intimacy.[99] The idea may even be repellent.

There's also maternal love,[100] the deep fundamental love a mother typically has for her child, which can be every bit as intense as romantic love, arguably even more so. We've covered how evolved brain processes and chemicals such as oxytocin give rise to powerful emotional bonds and motivations, and how these attachments are a key factor in a child's development.[101]

These and other more subtle variations of love clearly have a significant impact on our lives and our emotional wellbeing. So what is it about romantic love that gives it such heft, and makes the loss of a partner so traumatic?

One obvious factor is physical attraction. We usually don't just romantically love a specific person, from afar, in the abstract; we lust after them too, something absent from the other types of love. Lust is the brain-based manifestation of our sex drive, our fundamental urge to mate and reproduce. It's where our subconscious brain processes go, 'I am sexually aroused by the qualities of this

---

* Or more. The human brain is perfectly capable of open relationships, polyamory, etc. Monogamy is typically the default, though.

person I'm observing and am thus keen to engage in physically intimate acts with them'.

Obviously, we don't think those exact words. In fact, when it comes to lust, we often don't think *at all*. Sexual arousal, particularly the physiological changes that occur in places like our genitals, can happen without the brain proper getting involved, via reflexive processes stemming from neurons in the spinal cord.[102]

Even so, the brain often plays a prominent role in sexual attraction and desire. When we experience lust, there's increased activity in areas like the amygdala, hippocampus, thalamus, and more.[103] These are all crucial areas for multiple brain functions, and all are heavily involved in the processing and experiencing of emotions. Some even argue that, rather than an 'urge' or 'drive' or something like that, lust should be classed as an emotional state in its own right.[104]

However, sex and sexual attraction is only one part of developing and experiencing romantic love for someone. We see sexually attractive people all the time; they're omnipresent in the modern media landscape. But we don't constantly fall in love with the alluring individuals we see on our screens.

In fact, sexual attraction isn't even an *essential* aspect of developing romantic love for someone. It can occur much later in the process. It's a common trope of sitcoms and romcoms, where two people end up falling in love despite having been friends/enemies for years, with zero sexual aspect to their relationship. The human brain is powerful and flexible enough not to have to get a jump start from raw physical attraction to end up loving someone.

It can go even further. There is growing societal awareness and acceptance that some people are asexual. As in, for whatever reason, some people have very little or zero sexual desire or lustful inclinations towards others.[105] While there is still much debate as to how asexuality happens and how it should be classed, what's telling is that asexual people still regularly form romantic relationships.[106] This strongly suggests that romantic love and lust are separate things, insofar as our brains are concerned.

Indeed, several brain-scanning studies support this conclusion. For instance, when someone experiences lust, there's a notable spike of activity in the anterior insular cortex. But when they experience romantic love, there's a spike of activity in the *posterior* insular cortex.[107] This may not sound like much; after all, what's one part of the insular cortex compared to another?

It's not that simple, though. The insular cortex has been mentioned already, because it is heavily involved in the production, perception, and sharing of emotions; it's a crucial region for processing emotions like disgust, and also a key area for the process of empathy. But research reveals that different parts of the insular cortex have different roles. Specifically, the anterior, or front, of the insular cortex, handles more self-focussed emotional experiences. But the further back, or 'posterior', you go, the more complex and abstract the emotional information becomes.[108]

Essentially, the front of the insular cortex is concerned with 'I like . . .', 'I want . . .', 'I feel . . .', while the back part is more 'I like this because . . .', 'I feel strong emotional objection to this because it means . . .', 'I feel this, and here are the reasons why . . .' It's sort of the neurological and emotional equivalent of the classic diagram showing the evolutionary ascent of man; at the front of the insular cortex, you get the hairy knuckle-walking chimp-like creature, and at the back you have modern-day humans, walking upright and swinging their briefcases.

So, lust causing more activity in the front part means it's a more instinctive, short-term, self-focussed sensation, while love is more complex, more abstract, more inclusive of the higher brain regions. To put it even more simply, it strongly suggests that lust is instinctive, while love is far more cognitive. Love requires more thought, hence it's tricky to fall in love instantly. Lust has no such limitations, hence the many people who have woken up next to someone they'd really rather not be waking up next to.

Of course, this isn't to say love is an entirely abstract, cognitive, higher brain phenomenon. Far from it. Love occurs thanks to many of our subconscious, emotional workings. For example, romantic love boosts the levels of dopamine in our brains.[109] Dopamine is often described as a 'happy chemical', as it's the neurotransmitter used by the reward pathway, the circuit deep within our brain that allows us to experience pleasure.[110] If love boosts dopamine in this part of the brain, then it's no wonder it feels so good, so euphoric.

However, as I've taken pains to point out whenever the opportunity arises, dopamine activity in the brain isn't *just* about the experience of reward and pleasure. It has numerous other important functions, involving self-control, cognition, motivation, and so on. Falling in love therefore affects *all* these things. It's no wonder the experience can be genuinely disorienting, and have such profound effects on our behaviour and thinking.

Another important neurological region in the study of romantic love is the caudate nucleus, a large component of the basal ganglia, that deep-brain region with many vital functions for our subconscious and emotions. Evidence suggests the caudate nucleus is responsible for approach-attachment behaviour,[111] which is where we recognise something as significant and beneficial, and are thus motivated to behave in ways that keep it close, or accessible (depending on what it is).

As explored in Chapter 2, one thing our brain is always doing as we interact with the world is saying, 'There's a thing I want, I'm going to do what is required to obtain it/interact with it'. This can be expressed at varying levels of complexity, from the very simple, like, 'I am thirsty, there is water, I'm going to go drink it', to sophisticated, like how babies and small children will form attachments and stick close to their mother/primary caregiver.[112]

It can also get *very* sophisticated, like when we fall romantically in love with someone. It's hard to think of a scenario where we're more motivated to seek out and be around someone (i.e., approach-attachment behaviour) than when we fall in love with them, especially in the early stages. Hence, the caudate nucleus consistently shows raised activity in the brains of people who are deeply in love, which explains why love isn't just an emotional sensation, but something that strongly motivates us to do things, to think and behave in ways that would enable and validate our love.[113]

Having said that, something else we saw back in Chapter 2 was that such emotion-powered drives and motivations don't have free rein in the brain. More often than not, our complex cognitive processes get involved. We humans aren't creatures of pure impulse like that, because, thankfully, our brains give us the gift of executive functioning, the ability to consciously control, modulate, even suppress our baser instincts. Presumably, if love were purely an emotional phenomenon, the smarter elements of our brain would be able to rein it in.

That's not the case, though. Love and its influences, and consequences, aren't confined to the emotional centres of our brain. It affects the more cognitive regions just as readily, to the extent that trying to intellectually control and limit your love for someone can be like trying to get a refund on a sandwich a week after you've eaten it: not technically impossible, but it's an uphill struggle, to say the least.

Among the higher, more cognitive parts of the brain, scans of people in love show elevated activity in the occipitotemporal and fusiform regions (at the back of the brain, in the occipital lobe), the angular gyrus, the dorsolateral middle frontal gyrus (in the frontal lobes), superior temporal gyrus (a key region for empathy and mentalising, like we've seen), and more.

If you want to skip the specifics, just be aware that these brain regions are involved in social cognition, attention, memory, mental associations, self-representations, and beyond. All sophisticated, cognitive 'higher brain' processes, and all of them are affected by love, hence love affects how we think, how we remember, our feelings and attitudes towards people and things, how we see ourselves, and more.

Other brain processes combining emotion and cognition that are suppressed or disrupted when we fall in love include mentalising, or cognitive empathy.[114] This explains why the person we love can typically do no wrong in our eyes; our loved-up brain's ability to question or assess the thinking and motives of our romantic partner is seriously compromised.* This, coupled with elevated positive emotions and suppressed negative emotions, means our feelings about our romantic partners are often 100 per cent positive.

And the longer we're with them, the more established our love can become, as all these positive emotional reactions we have to our partner directly enhance our memories of the time we share with them.[115] So, the person we love becomes more prominent and enduring in our memory (thanks to the fading affect bias) than less emotionally evocative experiences and individuals.

In short, loving someone really is a big deal for our brains and emotions. It's no wonder it's so distressing when a romantic relationship fails. But then, given how the brain seems to go all in when falling in love, why is it that romantic relationships fail *at all*, let alone so frequently? How come we fall *out* of love with people, as well as in?

There are many factors to consider. For one, people aren't static. If you fall in love with someone, it doesn't mean they're suddenly

---

* A similar process happens with maternal love, hence many parents will insist that their child is an adorable little angel, even while watching them have a café-trashing public tantrum because they were given the wrong colour juice.

preserved in amber for all time. People change as they grow, age, and their circumstances alter. The person you love could be a radically different individual after ten years together, and all the experiences that involves. Life doesn't stop just because you're romantically involved with someone, no matter how many fairy stories or romcoms end here.

However, this is more of an external process, given how we're shaped by the world around us. In terms of what goes on in the brain, a major factor in maintaining a loving connection long term is, predictably, our emotions: how they manifest, and how we deal with them, in the context of our relationship.

Case in point: people often talk about 'the seven-year itch' or having 'lost the spark' in a relationship. There are countless stories and songs written about this. Basically, that romantic love fades or fizzles out is not something people seem to dispute. And when you know how the brain works at the most basic levels, this makes a grim kind of sense.

Particularly in the early stages, love is very demanding, in terms of energy and resources. We've seen how frugal the brain can be, so it would be reasonable to conclude that the brain can't sustain the initial throes of passion indefinitely. In fact, it has mechanisms in place to ensure it doesn't have to, like habituation,[116] which stop us responding strongly to anything that becomes too familiar, reserving our finite resources for dealing with the new and unexpected. And what could be more familiar than the person we spend every day with, for years on end?

Also, it's often said that being in love is like being on drugs (both activate very similar parts of the brain, admittedly[117]), but a brain constantly exposed to drugs develops tolerance; our flexible neurological systems alter and adjust to compensate for the presence of the new chemical, so as to restore some form of normal functioning. Given how neurologically disruptive it is, our brains would pretty much *have* to do the same with love.

Is that why love fades over time? Because our brains just get used to our romantic partner, and learn to tune out the emotions they once provoked in us? It's a bleak thought, and certainly doesn't support the notion of 'happily ever after'. Thankfully, while these processes no doubt play their part, the brain has many more tricks up its sleeve to sustain a relationship.

Habituation, while a fundamental cell-level process, usually doesn't apply to things that our brain regards as 'biologically significant', like food.[118] We may tire of a specific *type* of food if we eat it too often, but we seldom stop enjoying eating in general. Given how deeply embedded love, lust, and the sex drive are in our brains, the person we love also qualifies as 'biologically significant'.

Also, subtle relationship changes after several years together can help counteract the process of our loved one becoming too familiar. You may feel you know everything about your partner. But do you know them as a pet owner? A manager? A homeowner? A parent? As they change and grow, you may end up loving them *more*.

Research suggests this belief that love eventually fades is wrong. There's no reason to assume that a couple will fall out of love purely due to the passage of time. Many long-term partners seem to be as in love with each other as newer couples still caught up in the throes of passion.[119] And 'passion' is the key word there. It appears the idea of love fading over time stems from people conflating romantic love with lust, or what's described by some as 'passionate love', the combination of romantic love and lustful yearnings.

It's *this* intense, demanding stage of love that our brain typically can't sustain for too long, and it's far more common in the early days of a romantic relationship, so it predictably will fade away if the relationship endures. And, despite what many a fictional portrayal suggests, science says it's a *good* thing this happens.

Research reveals that passion and lust are considered to be positive qualities in a partner by those in the early stages of a relationship. By contrast, they're often considered to be *negative* things by those who've been with a partner for many years.[120] We can chalk part of this up to the fact that people in longer-term relationships are typically older. Sex is a physically demanding act, and when we're in our youthful hormone-infused prime, we want to do it as much as possible. But when your body has aged somewhat, it both lacks the energy and endurance it once had for a rigorous sex life, and isn't as strongly influenced by the hormones that sustain such a thing.[121]

It goes beyond the physical and hormonal challenges, too. We've seen that falling in love has a lot of positive, if disruptive, emotional effects on our brain. But they're not *all* positive. When we fall for someone, and fall hard, we don't just love them; we become obsessed with them. And this can stir up a lot of negative emotions: paranoia that they'll leave us for someone else; jealousy of anyone else they interact with; desire to 'protect' them, which can manifest as trying to control and restrict them, to keep them to yourself. And so on.[122]

If both partners in a relationship feel the same way about each other, these things aren't such a problem. Many a new couple has seemingly vanished from their friends' lives as they spend every waking moment together. But if there's any disparity, like if one partner gets over the obsession phase before the other and wants to resume other aspects of their life that don't necessarily include them, an obsessed and controlling romantic partner can quickly become restrictive and suffocating. You'd think, in this situation, the obsessed partner would see they were upsetting the person they love, and change their behaviour. But remember, being in love suppresses our ability to do this, to gain insight into what our partner is thinking. Nobody said that was always a good thing.

Then there's the data which strongly suggests the emotional connection between two romantic partners is all important, even more so than the physical. This echoes what we saw about the appeal of BDSM. Maybe that's why people say 'love hurts'?

In any case, one study investigated attitudes towards infidelity, asking which one people found worse: sexual infidelity (your partner having sex with someone else) or emotional infidelity (where your partner forms an emotional bond with someone outside your relationship, which excludes you).[123] Female subjects largely considered emotional infidelity worse than sexual infidelity, which arguably conforms to gender stereotypes somewhat. But the shallower, emotionally reserved, sex-and-status-obsessed men? They *also* felt that emotional infidelity was worse than sexual. Once again, it seems men and women are more similar than not, at least with regards to emotions.

This may also explain why some people can happily be part of open relationships, where they are *physically* intimate with others, but only *emotionally* involved with their partner. Or why people stay together long after their sex life has declined. The main theme here is that, in romantic relationships, emotional connections often trump the physical. And it's this emotional connection, how you emotionally communicate and engage with your partner, that can often make or break a relationship.

Every couple will have disagreements and disputes. It's inevitable. What's also inevitable is that arguing with someone you love will result in negative emotions. It's what we do with these, how we regulate them, that can often determine whether the relationship survives them.

A 2003 study into this[124] revealed that many people in relationships opt to suppress, or deny, the negative emotions they experience following disputes with their partner. While this may be an effective short-term solution for maintaining the status quo of a romantic relationship, and avoiding further emotion-fuelled conflict, it's very unhelpful. The study found that the long-term result of emotional suppression is that people tend to forget the specific details of what was said, or what the dispute was about, and instead remember the negative emotions experienced. So, you're left feeling upset and resentful, while the cause of the dispute (and subsequently the negative emotions) is left unaddressed and poorly remembered, drastically increasing the chances of it happening again.

The upshot is, if you repeatedly argue with your partner, but suppress the resultant emotions, these feelings build up, because by not engaging with them, you don't effectively process them. Eventually you'll have a great many negative emotional associations with your partner, with no obvious source or cause. Could these eventually overrule your love for them? Quite possibly, yes.

By contrast, the same study showed that if you reappraise your emotional reactions to disputes, you end up remembering the details and specifics of the argument, but not so much the negative emotions they provoked. And you reappraise your emotions by reconsidering them, reinterpreting them. It's said that couples should talk through their problems, and that's what allows this to happen.

Say your partner forgets the anniversary of your first date. This causes you to feel anger, and sadness. Suppressing these emotions means they just sit there in your brain, impacting on everything else it's doing. You know when someone's clearly frustrated or upset but, when asked, they actively deny it and insist they're 'fine'? This is classic emotional suppression.*

However, what if you admit to your partner that you're angry about the forgotten anniversary, and they explain that they didn't forget, but the gift they ordered hasn't arrived due to postal delays?

* It's also another good demonstration of empathy because, despite their objections and efforts, it's obvious that they aren't 'fine'.

Or that they forgot because they were distracted by organising the holiday you've been looking forward to? Or that they simply didn't realise this anniversary meant so much to you, but that they'll do better in future?

In each case, more information has been obtained, and your emotional reaction to the anniversary forgetting can be adjusted accordingly. You can go from anger to happiness or satisfaction because they didn't forget, or were distracted by something even better.

Granted, you might be annoyed if they simply didn't appreciate the importance of the anniversary, but that's still much better than unspecified anger. We've already seen that our brain is constantly reassessing and updating our emotional responses, based on new experiences and information.[125] Talking through the incident, emotionally reappraising the dispute, allows this process to happen. By contrast, suppressing the anger and sadness means you're stuck with those feelings, and remember them for longer, without remembering exactly why or how your partner hurt you. You just know they did.

This process isn't limited to arguments and disputes. Research suggests that our long-term romantic partners often become something of an emotional 'modulator' to us. They allow us to better experience, process, and control our emotions, just by being part of our lives.[126]

Have you ever felt frustrated with your partner because whenever you tell them about something that's bothering you, their immediate reaction is to offer potential solutions, or try to 'fix' things for you? At face value, this seems a valid approach: you admit to having a problem, and the person who loves you tries to help you with it. What's wrong with that?

But when you consider the importance of emotional communication within a relationship, reacting with frustration actually *does* make sense. When we tell our partner about something that's upsetting us, be it an infuriating work situation or dismay at our own poor progress at the gym or whatever, we're not necessarily looking for a solution or fix. What we're hoping for is to be able to express our emotions, in a safe context, and have them validated or empathised with. Having our partner listen to and support our emotional reactions allows us to better process them, and encourages us to keep having them, which is by far the most healthy and helpful approach.

On the other hand, if they try to offer solutions and fixes, however well meant, it can often feel like they're saying our emotions *aren't* valid. And if they're offering fixes that we already considered and dismissed, it can seem like they're dismissing our emotions and our intellect. Why wouldn't you react negatively to that?

This explains why studies going back decades have found that strong and enduring emotional connections and communication, via empathy, mentalising, and related processes, are a key factor in the maintenance, endurance of, and happiness within, romantic relationships.[127] Falling in love is one thing, but it seems like if you form, and maintain, strong and communicative emotional bonds, you get to *stay* in love.[128]

In the strongest romantic relationships, it could be that each partner becomes an important aspect of the other's emotional makeup. Have you ever enjoyed something, like a style of music or a certain TV show, purely because your partner likes it? It's a common occurrence, but one which is a clear demonstration that your partner has fundamentally altered your emotional responses. This may be part of the reason why people in long-term relationships tend to have better wellbeing, both mental and physical;[129] our emotions are an important factor in both, and those with a romantic partner have a notable advantage when it comes to emotional experience and processing.

Does this seem a bit much, though? Being physically intimate with someone is one thing, but having another person become such a significant factor in your emotional workings, that's something else. Our emotions are a crucial underlying component of pretty much everything our minds do. For our partners to play such a significant role in them would suggest they've become a factor in our very identity, our sense of self.

In fact, numerous studies suggest that's exactly what's happening. We've seen that multiple regions of the brain are activated when we experience love, but the level of activation often depends on how long the subject has been in love with their partner. However, there seems to be one particular neurological region that is activated when we're in love, and which stays active regardless of how long the relationship has been going on. And that's the angular gyrus.[130]

This is significant because the angular gyrus is strongly linked to (among other things) self-awareness, our sense of our own identity.[131] One obvious conclusion to be drawn from this data is that the person we love romantically, and our relationship with them, literally become a part of our very identity. And when you consider the effect of our romantic partner on our emotions, the ever-increasing percentage of our memories that include or are about them, and how all our plans and goals and ambitions now involve them, it would perhaps be more surprising if the person we love *didn't* become a significant component of who we are.

And this, I feel, truly explains why the breakdown of a long-term romantic relationship, the loss of a romantic partner, is so emotionally devastating. There are all the plans and expectations now rendered invalid, all the happy memories soured, all the emotional investment squandered, all the stress of a future that's suddenly way more uncertain, all the practical demands and requirements involved in separating two lives that were previously intertwined. On top of that, though, there's something even more fundamental and disruptive to consider: over time, the person we love and spend our lives with genuinely becomes a part of us, our very sense of self. And when they're gone, it feels like a part of us has been lost with them.

And this brings me back to my initial question: why does the loss of a romantic partner seemingly cause more emotional distress than the loss of a parent? After all, don't our parents also shape our emotions and identities? Don't they feature in the bulk of our memories? Don't we depend on them just as much? Yes, we do. When we're children.

And that, I'd argue, is the crux of the matter. During childhood, our parents play a massive part in our emotional existence and understanding of the world.[132] That's why the version of the Holmes and Rahe stress scale for children does indeed have 'loss of a parent' as the most stressful possible experience. But when we hit adolescence, we invariably put a great deal of time and effort into establishing our independence, our autonomy, our own identity. It's an important part of the human maturation process. And it invariably means we strive to become someone distinct from our parents' influences, something long recognised as an underlying cause of parent–teen conflict.[133]

In most cases the conflict eventually ends when we become fully mature, independent adults. This often means the relationship between parents and their adult children is more equal in status than during their childhood, when the parents are undoubtedly in charge.[134] Indeed, in time, the grown children may become the dominant ones in the relationship, if the parent becomes too old or infirm to look after themselves.

In many ways, the reverse happens with romantic partners. We start off as an independent individual adult, and we actively seek out someone to love. And if we find them, and the love is reciprocated, we spend significant time and effort in building emotional bonds with them, integrating them into our lives, our memories, our sense of self, often with the hope of becoming parents ourselves, and so the cycle, and therefore the species, continues.

I can't speak for everyone, obviously. Everyone has distinct and unique relationships with their parents. Some will remain incredibly close during adulthood, others will become increasingly estranged, and everything in between. I can only give my own perspective on this.

The fact is, I loved my dad. It was a bizarre sort of love, but it was there. However, even before he died, it had been a long time since I've felt I *needed* him. As a result, losing him, as much as it hurt, was something I could, apparently, cope with.

The same absolutely cannot be said for my wife. I love her, and during these difficult times I've needed her more than ever. And if I lost her, I honestly don't know what I'd do.

Love thy stranger: how and why we form one-sided emotional relationships

My father was a popular man. During my youth, he was a pillar of the local community, what with being landlord of the local pub. That never changed as we grew older; Dad was invariably the life and soul of the party (and often responsible for throwing said party). Of course, this meant that a great many people were deeply saddened by his passing, all of whom would have ideally been present at his funeral.

Sadly, the funeral occurred under strict lockdown rules, and only fourteen people were allowed to attend. This meant we ended up having hundreds of people lining the streets of Port Talbot (where Dad lived) to see the funeral procession pass, and pay their respects that way. And that's just those who could do so under strict travel restrictions. It could have been twice as many again.

It was a profound experience, sitting in the funeral car, following the hearse conveying my father's coffin, as we slowly passed dozens of sombre, black-garbed people, many of whom were figures from my childhood, or extended family, that I'd not seen in years. All set against the backdrop of suburban Welsh streets with rolling hills behind, the vista of my youth.

Unfortunately, given my (at the time) incompetence at recognising and dealing with my emotions, I resorted to feeble humour, and observed that it all felt like a Welsh re-enactment of Princess Diana's funeral. However, this glib comparison stuck with me for a long time afterwards, in a way that seemed even less funny than it did originally.

The death of Princess Diana in 1997 affected millions of people across the world. My father's passing, as devastating as it was for myself and many others, could never hope to generate an iota of the grief that Diana's did. Clearly, many people felt great affection, love even, for her. Yet, with the exception of an infinitesimally small percentage, none of these people *knew* her. Not personally. All they knew about Princess Diana was gleaned indirectly, via media coverage with varying degrees of accuracy and morals. And the reverse is even truer: Princess Diana was unaware of the existence of most of the individuals who loved her.

'Dunbar's number' is the theory, first proposed by UK anthropologist and evolutionary psychologist Robin Dunbar, that there is a maximum number of stable social relationships we can form and maintain, due to the nature and makeup of the human brain. And that number is 150: Dunbar's number.[135]

There are numerous observations and counterarguments that question the validity of Dunbar's number.[136] For instance, some people struggle to keep in touch with a couple of dozen friends at most, while others will maintain strong connections with well over 150 people (my father was certainly one of these). Every brain is unique, after all.

But, even taking all that into account, the notion that the human brain can only sustain a limited number of relationships makes logical sense, because when we form a social connection with someone, it involves a considerable emotional component. Our cognition can go, 'We have qualities in common with this person', but it's our emotions that say, 'We *like* this person, we enjoy their presence, and thus want to be around them more'.

So, given that each meaningful social relationship requires significant emotional resources, our typically frugal brains would logically have an upper limit on how many they can sustain. Some say we have an upper limit on friendships because we lack the 'emotional bandwidth' for more.[137] Impressive as it is, the human brain still has limitations.

It's not *all* emotional; maintaining and nurturing social relationships requires a lot of cognitive work, so the rational, intellectual elements of our brains are also heavily involved. Imagine you're chatting with a friend, and you see an opportunity to make a funny yet tasteless joke, but spend a brief moment figuring out whether your friend will laugh or be horrified before you say it. Your brain essentially uses all the information you have about your friend to run a simulation of how they'll respond.

But consider all the information, and manipulation thereof, that's required to do this, and to do this in fractions of a second. It's a considerable feat of cognitive power, which we perform all the time, during any interaction. As is the case with emotions, this cognitive workload uses up the brain's reserves of energy and resources.[138] Where our friends are concerned, our brains have a wealth of information and emotion to work with. We've seen what falling in love does to the brain, and that friendship is technically another form of love. Is it any wonder that social engagements, no matter how enjoyable and rewarding they may be, can also be very draining?[139]

So, given all this, isn't it a bit surprising that uncountable millions of people regularly put considerable time and emotional investment into individuals that they'll likely never even meet, let alone form a meaningful connection with? In fact, when you look at *Star Wars* fandom, *Harry Potter* obsessives, 'Bronies',* all the passionate followers and admirers of various video game and anime characters, and more, it shows that countless people dedicate a great deal of their 'emotional bandwidth' to individuals, characters, and places that do not, and often *cannot*, exist in reality. How?

Well, thanks to modern mass media, we are now regularly exposed to other people's thoughts, views, personal lives, appearance, clothing, sense of humour, creative endeavours, conversations, and more, without ever being in the same room as them. Or even the same country. And this is where it gets interesting.

Say you stumble upon a podcast about an area that interests you. You listen, and find the host engaging and amusing, their content informative and understandable. You learn their name and background and empathise with them when they speak about problems they're dealing with. Basically, you *like* them. It's just like when you meet and 'click' with someone in a standard social interaction.

Except this *wasn't* a standard social interaction. The podcast host has no knowledge of this exchange whatsoever. This a parasocial interaction.[140] It describes when you're exposed to another person in a manner other than face to face, are emotionally stimulated and engaged by them, but they remain blissfully unaware of the whole thing.

Then, say you're so emotionally engaged by the podcast you heard that you're motivated to seek out more. So, you subscribe to the podcast, listen to all existing episodes, look up other things the host has done, and so on. This is now a parasocial *relationship*;[141] the

* Adult males who are die-hard fans of My Little Pony. Yes, it's a thing.

parasocial interactions are ongoing, and you invest mental and emotional energy into the liked individual, just like a close relationship in the real world. But the person you're investing your emotional energies into remains unaware and uninvolved in the process.

Now, when you consider how important real emotional bonds evidently are, and how neurologically demanding such things can be, parasocial relationships can seem unlikely, unhelpful, and straight up weird. But when you delve deeper into the underlying science, it makes a lot more sense.

Parasocial relationships may not be 'real' in the strict objective sense but, for our brains, telling the difference between what's real and what's not isn't as straightforward as you'd think. Whether it's information obtained via our senses as they engage with the world around us (i.e. reality), or the emotions, thoughts, memories, and predictions constantly being conjured up internally, it all ends up represented as patterns of activity in, and between, neurons. So, in a sense, it all looks the same, as far as the brain is concerned.

Thankfully, our brains have systems in place to keep sensory information from the real world separate from the information it produces itself, via its own internal processes. These systems involve multiple important brain regions, including the thalamus for handling raw sensory data, the sensory cortex and associated connections for converting it into perception, the hippocampus for memory encoding and retrieval, the frontal lobes for consciously recognising and utilising information, and more.[142]

Unfortunately, as impressive and sophisticated as this system may be, it's not 100 per cent reliable. Disruptions in this network are an underlying factor in conditions like schizophrenia, which commonly feature hallucinations and delusions, both of which are manifestations of a brain perceiving internally generated phenomena as if they're 'real'.[143]

But even in a typical healthy brain, the line between 'real' and 'not real' is a surprisingly blurry one. For example, when we recall or imagine something, we often use mental imagery. We visualise it, in our so-called 'mind's eye'. Studies show that doing this leads to activation of the brain's visual systems, just like seeing something with our eyeballs does.[144] And it's by no means only the visual cortex that is activated by both real and imaginary things; far from being solely about idle fantasies or artistic expression, our imagination is a fundamentally important part of how the human brain operates.

A significant part of our mental existence involves predicting things, anticipating outcomes, dwelling on possibilities, making long-term plans, forming ambitions and goals, navigating around unfamiliar locations, and so on. Whenever we do any of these things, our brain is generating simulations: mental representations of scenarios, situations, outcomes, locations, even individuals, in forms that currently do not exist or have not occurred, and may never do. So, far from being a distraction or irrelevance, imagination is a crucial element of our cognition, our ability to interact with the world and function within it.

The neurological basis for imagination reinforces this point. Various studies have yielded data suggesting that imagination and prediction overlap a lot with memory.[145] Remember, when we recall a memory, it's believed to be the result of our brains rapidly reconstructing it from the relevant elements of that particular memory, stored separately throughout our cortex, but activated in the correct formation.[146]

However, one thing the brain is known to do is take processes it's evolved to fulfil one purpose, and adapt them for other uses, like we saw earlier with oxytocin. So, if memory recall is our brain activating stored information in the *correct* pattern, what's stopping our brain activating information in an *incorrect* pattern? Nothing, really. It happens often.

We all have memories that we recall vividly but which have aspects that are wrong in some way: we remember a certain event but misremember who we were there with, we attribute a well-known saying to the wrong person or source, we disagree with others about what happened at an event you were all present at, meaning that someone's memory must be wrong. Indeed, Chapter 3 argued that dreams are essentially this: sequences of memory elements, being activated randomly and out of context, hence they're so weird.

Imagination is the conscious, deliberate version of this, which allows us to construct mental simulations of potential events and experiences, deduce their outcomes, and use this information to our benefit. In light of this, it makes sense that the hippocampus seems a key area for imagination and predictions, not just memory. This is according to multiple studies,[147] as well as data revealing that individuals with damage to the hippocampus struggle to imagine things or envisage future scenarios.[148]

Again, there's a logic to this: while the things we imagine usually haven't happened and likely never will, the specific details of the things we imagine, be they person, place, or events, are invariably things we've encountered before, and committed to memory. It's very difficult to imagine something entirely unique, with character-istics and qualities that we've no prior experience of. It's like trying to imagine a brand-new colour or shape: practically impossible. The memories stored in our brain are to our imagination what the letters of the alphabet are to novels: they can be combined and expressed in a vast number of different ways, leading to almost infinite creativity.

But you can't write stories with letters that don't exist, that nobody recognises. Therefore, it makes sense that the hippocampus, the hub of memory in the brain, responsible for storing, retrieving, and linking information together, would be an integral part of the imagination process.

Although, it's not *just* the hippocampus that's responsible for imagination and all that it can do. That would be quite a challenge for any single neurological region. No, there's also what's been dubbed a 'core network', a circuit encompassing multiple brain regions, including (but not limited to) the medial and lateral prefrontal cortex, posterior cingulate cortex, retrosplenial cortex, lateral temporal cortex, and the medial temporal lobes.[149] These areas are spread over much of the brain, and the way they work with the hippocampus to provide the various forms and uses of imagination we're capable of is baffling, and still being explored. Suffice to say, this core network, including so many important regions for both cognitive and emotional processes, further emphasises just how fundamentally important our imagination truly is to how we operate.

This doesn't just apply to the decisions we make, or other complex cognitive activities like that. The things we imagine can influence and shape what our brain does on a more direct, fundamental level. Using our imagination to anticipate or predict something can end up altering what our brain perceives when we're exposed to it for real. Studies have shown that if we predict someone won't have a strong reaction to something, we tend to perceive their eventual reaction as less potent than it is.[150] Other studies show that if we anticipate a smell will be bad, we'll find it unpleasant when we eventually smell it for real, even if it's a perfectly neutral odour.[151] Imagining something happening can even cause tangible reactions, subconscious physical reflexes in our body, like pupil diameter changes.[152]

The takeaway from all this is that, despite technically not existing in the real world, the things we imagine, fantasise, and hypothesise about, can have real, tangible impacts on our brains and bodies. Emotions are a key factor in this. Many of the regions involved in imagination and prediction are ones which also play prominent roles in our emotions, and our imagination can trigger genuine emotional reactions, too. This is pretty much the basis for the horror industry. The sense of 'dread', which so much horror-based media goes to great lengths to induce, is a persistent feeling of fear regarding what *might* happen, rather than what we're actually seeing/hearing/reading there and then. And this is entirely dependent on our imagination-based anticipation.

It doesn't even need to be something actively designed to scare us. So much modern-day stress is caused by us being afraid of or worried about things that *may* happen. What if we lose our job? What if our partner leaves us? What if the political party we don't like gets into power? What if we miss our flight? While these are all sensible, rational things to worry about, the fact is that, in the vast majority of cases, they *haven't* occurred, at least not yet. They may never do. But we *imagine* them doing so, we simulate such outcomes, and experience genuine, visceral, physically impactful emotional responses to them.

Here's what I'm getting at: if the things we imagine, that exist entirely as a simulated construct within the confines of our own brain, can trigger genuine fear, what's stopping them from triggering other emotions, and emotional processes, too? Like happiness, love, affection, empathy?

Nothing, that's what. Hence people readily form strong emotional bonds with individuals they've never met, or that don't even exist in the real world. As long as they have enough information to create a simulation of someone in their heads,* then our brain's ability to extrapolate and imagine seems easily powerful enough to result in this simulation inducing genuine emotional investment. All told, it's no wonder that the same neurological mechanisms are

---

* And our technology-infused omnipresent media landscape provides this in spades.

used for both social and parasocial relationships.[153]

So readily does the brain form emotional connections with individuals that we'll never meet/don't exist, there are multiple ways this can occur. In addition to parasocial interactions and relationships, another process is known as transportation.[154] Have you ever experienced being 'lost' in a good book, TV show, film, or video game? It's where we become so emotionally engaged and engrossed in something, we 'tune out' the world around us to focus on the fictional one we're absorbed by. We're transported from the real world to a fantasy one, insofar as our conscious mind is concerned.

This may be yet another example of our brain's limited ability to do multiple things at once. If a story is especially absorbing, the plot gripping, and the characters likeable and compelling, then our highly stimulated emotional systems will direct more neurological resources into exploring it. Which means fewer resources available for paying attention to the actual world around us.

This transportation phenomenon doesn't happen with any old fictional entertainment, of course. For one, it's widely agreed that it requires some sort of narrative in the media being consumed, whether it's fiction or non-fiction. A narrative is important for how our brain understands things/events, for various complex reasons.[155] It helps us figure out the relationship between the people and the happenings being described, and the world they inhabit; it provides a structure and pattern otherwise absent in abstract information; it involves change and dynamism which we're inherently more attentive to; and so on.

But as well as narrative, the transportation effect requires *characters*.[156] We need someone we can relate to, understand, and empathise with. It's all well and good describing a series of impressive and important events in a place or time other than your own, but however significant they may have been, it can often seem too abstract or distant to engage with for the average listener.* Very few compelling stories are simply descriptions of important things that happened.

If you include relatable characters then you have an 'in' to the story. Thanks to how our brains work, it's much easier to emotionally connect and relate to another thinking, feeling individual, rather than events, environments, or situations. It's as if the characters in stories act as something of a conduit for, or translator of, the wider narrative context for our emotional processes.

Another process at work when we become emotionally invested in distant or fictional individuals is identification.[157] Much of the time, when we become emotionally invested in a famous or admirable person, it's because we identify with them; we think they're like us, or we want to be more like them. Given how focussed humans are on their social status, combined with how we're always instinctively learning from others, it's perfectly normal for people to see the qualities of higher-status individuals, and try to obtain these qualities themselves, by emulating the individual they admire.

Some might scoff at this sort of thing, thinking it's beneath them, but identifying with prominent individuals in this manner has been going on for as long as our civilisation has.[158]

Studies suggest we tend to enjoy media and entertainments more when they involve characters we identify with. We find them easier to empathise with, so become more invested in their story.[159] That we're so influenced by the individual we admire, and so keen to identify with them, is why celebrity endorsements can be so powerful and effective. If someone we're deeply invested in relays a certain message, endorses a certain product, or is seen wearing a certain designer garment, we're far more likely to agree with that

---

* As anyone who's ever been bored during a particularly dry history lesson in school will be very aware of.

message, or buy that product/garment.[160]

But it's not all about using famous faces to sell tat to the masses. On the flip side, our emotional involvement with prominent individuals can also be used to help spread helpful information: to influence, educate, and inform us in a positive manner.

This effect can be very potent for children. One particularly interesting study showed that US toddlers learn mathematical skills better if taught them by a familiar character they like (in this case, *Sesame Street*'s Elmo) than if taught by one they don't know (in this case, Dodo, a Taiwanese character unfamiliar to most US toddlers). However, if the toddlers were allowed to 'emotionally engage' with the more obscure character Dodo, by playing with tie-in toys and watching their TV show, Dodo's educational impact on them quickly rose to match that of Elmo's.[161]

A similar study showed that toddlers can learn new skills from interactive media characters that they can engage and play with, but that they learn a lot better when the character is tailored to them. If the character uses the toddler's name and has qualities that the toddler likes (e.g. being their favourite colour), then they're a far more effective teaching tool than a generic character that isn't tailored to the specific toddler.[162]

This makes sense when we consider that emotions help us work out what things are to be accepted/pursued, and what are to be avoided. Hence, when individuals we're emotionally positive about impart information, we're predisposed to accept it more readily. The inverse is also true: if you've a negative emotional connection to someone, any information they impart to you will be hard to accept. If you've ever struggled with a subject at school because you really don't like the teacher, then you'll know all about this.

As adults, too, emotional investment in individuals who can/ will never reciprocate can be very helpful for our development and mental wellbeing, despite the many negative portrayals and stigma.*

While emotionally rewarding relationships with real individuals will always be the ideal outcome, some people struggle with this. It may be due to social anxieties like shyness, circumstances like living in a remote location where nobody's around to befriend, or something else entirely. Whatever the cause, it happens, and surprisingly often.[163] However, much evidence suggests that, if you struggle with the more tangible options, parasocial relationships can be good for your wellbeing, motivation, and more.[164] They may not be technically real, but they are real *enough*, at least as far as some parts of our brain are concerned.

Parasocial relationships can also be educational and informative in their own right. I bet we've all rehearsed conversations in our head before an important interview. Such discussions don't exist outside the confines of our mind, but they allow us to practise and refine our interactions, nonetheless. Similarly, we've all spent hours mentally replaying arguments, figuring out what we *should* have said, or *will* say next time you see your antagonist. It's the same thing: using imaginary interactions (albeit ones derived from real-world information) to refine and rehearse your responses, should a relevant situation genuinely occur in the future.

As children, imaginary friends are parasocial relationships in which the fictional character is wholly created by the relationship instigator. Children have potent imaginations, a drive to peer bond and interact with others, and a strong urge to play.[165] At the same time, their brains have much less experience of how the real world operates, so would have a harder time differentiating between what's actually happening and what's created by their own minds.

---

* It's very common to see enthusiastic fans of anything portrayed in mocking or derogatory ways, as 'nerds' or 'loners' who should 'get a life', and so on.

It's therefore unsurprising that some children would construct a simulation of an individual using their imagination, but go so far as to emotionally bond with said simulation, and behave towards it as if it were a real, tangible friend.

Many parents may feel that their child having an imaginary friend is something to be concerned about. But the available evidence shows that they can have many benefits: they act as sources of comfort in times of boredom and loneliness, as mentors for children in their academic pursuits, and they're often encouraging, motivational, and good for self-esteem. They can even serve as moral guides for children, sort of acting as an interactive conscience, helping them think through and make the right moral decision.[166]

Children with imaginary friends are often developmentally ahead of others too, in terms of language and social interaction.[167] It's as if they've had a lot more practice with communicating and engaging with someone else than the average child has. Imagine that!

The benefits of parasocial relationships occur in the later stages of youth too. Adolescence is a tricky and confusing time for everyone, for various reasons, and one well-known frustrating aspect of our teenage years is the tendency to form 'crushes', where we become almost helplessly infatuated with someone.[168]

Our crushes can be on anyone we find desirable,* be they a celebrity, fictional character, or classmate. What can make them so frustrating, though, is that the 'crushee' is invariably unaware of the feelings of the crush holder, and the effect they're having on them. Therefore, crushes are yet another form of parasocial relationship.

---

* Much of the literature focuses on crushes in teenage girls specifically, even though teen boys are also prone to them. I don't know if this is because of differences in the male and female maturation process, or if it's just another example of 'women = beholden to their emotions' stereotyping creeping in again. Could feasibly be both.

However, despite all the stress, distraction, frustrated desire, and inability to do anything about it that adolescent crushes often lead to, research once again suggests they're an important and useful aspect of our maturation. Statistically, having an intense adolescent crush often leads to increased chances of finding and experiencing fulfilling romantic love later in life, and bolsters confidence in a relationship.[169] And it's not just the romantic aspect, as teen crushes often have a hormone-fuelled, erotic, sexual element too. Even this, and other sexual fantasising, can aid development, and our ability to handle and process sexual interactions later in life.[170] The general gist still seems to be that if we can fantasise about it, we can learn from it – and do so without the risk of real-life heartbreak inherent in a romantic relationship.

So, far from being a waste of the brain's limited resources, parasocial relationships actually have many important uses. They allow our brains to figure things out and develop in safe, low-risk ways. They can improve wellbeing. They help us learn and absorb information. They shape how we see ourselves, and motivate us to improve by giving us examples to aspire to and emulate. And much of the time, they just make us happy. That's often helpful enough.

Having said all this, I'd be remiss if I didn't flag up that, as is often the case, there are down-sides to parasocial relationships too.

Occasionally, people *do* get to meet the individuals they have an emotionally potent parasocial relationship with. And while it may be a gleeful, once-in-a-lifetime experience for the fan, it can be disconcerting and baffling for the individual on the receiving end.

I've previously mentioned *Star Trek*'s Lieutenant Commander Data, who is a very popular character, beloved by many fans of the franchise. Of course, Data isn't a real person. He's a fictional character, played by actor Brent Spiner, who is well acquainted with encountering superfans. Thanks to an obscenely unlikely sequence of occurrences, Spiner and I happen to have friends in common. This meant I was able to ask him what it's like to meet people who are so emotionally invested in a character he played:

> It just happens that, for many people, Data is the only reference for me that they have. And as pleased as I am for their attention, I can't say I fully enjoy being referred to by his name.
>
> Often when people write me or see me in public, they call me Data. 'Hey Data!', and so on. I know they mean nothing but the best. But it seems to erase the rest of my life. As much as they would like me to be Data, in fact, in my mind, I am me. Data was, without question, a fantastic job with a multitude of fringe benefits. But even though it was a very good part, it was just one part of the total of my existence.
>
> Fans often think, when I say this sort of thing, that I hate the character. Far from it. I love Data. But love is a very complex emotion. And my relationship to Data is very different from anyone else's.

Spiner raises several interesting points there. For one, he had his own parasocial relationship with the character of Data, one very different to that of those who got to just observe him. And obviously, these two types of relationship won't gel well. I imagine it's like meeting someone who's fallen in love with your annoying sibling: they can't stop talking about how wonderful and amazing they are, and you can't help thinking of them as an infuriating self-centred brat.

As Spiner astutely observed, some studies have revealed that a parasocial relationship will 'overrule' the dissonance experienced when you see an actor in a different role than the character you associate them with.[171] If the fan is sufficiently emotionally invested in the parasocial relationship to want to 'protect' the object of their feelings, they may opt to assume the actor is still the character they've grown to love, not a wholly distinct human being with a life of their own.

While parasocial relationships can be beneficial for the person experiencing them, they can be confusing, even distressing, for those on the receiving end, should they ever be made aware of them. And it goes beyond just awkward conversations: it's become depressingly common to hear about the latest fan-led backlash against changes, alterations, or even just the ongoing development of well-known characters or fictional universes.[172] Such obsessive fans often end up sending death threats to those they feel responsible for 'ruining' the thing they love. It's pretty damning of the whole concept of fandom, and emphasises the high levels of emotion involved.

It's worse again when this unhealthy level of emotional investment is applied to a parasocial relationship with someone who's an actual person in the real world. While comparatively rare, there are those who end up experiencing celebrity worship, a phenomenon whereby someone's love for a famous person becomes all-consuming and takes over much of their life.

Because parasocial relationships exist entirely within our minds, we technically should have total control over them. But often, that's not how it works; the individuals or characters we've dedicated ourselves to invariably have their own independent existence, which we have no say in or influence over. This can be very frustrating, to be so emotionally invested in someone, to so want to be with them and protect them, but to be regularly reminded that such things are forever beyond you. A loss of autonomy is stressful for the brain at the best of times.[173] In extreme cases, people may even feel compelled to exert 'control' over the relationship, inserting themselves into the life of the source of their obsession. This is how you end up with celebrity stalkers.[174]

Ultimately, the relationship we create in our heads typically has minimal chance of surviving contact with the real world. The people we admire are *not* our friends, because they don't know us. At *all*. Unfortunately, as we've seen, people may sometimes be so emotionally invested in the parasocial relationship they've created that their brains *overrule* the evidence of their senses. This would explain why unpleasant revelations about very popular figures regularly lead to immediate and forceful denial from their most enthusiastic fans. It would also explain why stalkers persist in their terrifying actions no matter how distressing and upsetting they are for the person they supposedly care so much about.

However, it's at least equally common that the evidence of our senses overrules the relationship we've created in our minds, which is why teenage crushes seldom survive meeting, and interacting with, the focus of said crush.[175] Because even if the crushee does nothing wrong whatsoever, it's virtually guaranteed they'll still be very different to the version of them that's been built up and refined in the mind of the crush holder. When presented with such conflicting data, it's much harder to sustain the fantasy. Doing so involves a hefty degree of denial, and most brains aren't so invested in parasocial relationships that they'll actively *block out reality* to sustain them.

In cases like this, ending a parasocial relationship is invariably the healthier, more rational option. Unfortunately, that's also not without its drawbacks. Research shows that when a parasocial relationship ends, for whatever reason or scandal, the emotional fallout can be very similar to that of a real-life relationship breakup.[176] People can experience intense grief (like with Princess Diana), or a powerful sense of betrayal.

Again, it makes sense: if our brains use the same processes and emotions in parasocial relationships as they do for genuine ones, then ending a parasocial relationship should have similar effects as a genuine breakup. The feelings may not be as intense, but then parasocial relationships rarely feel *exactly* as rewarding and fulfilling as the real thing, so the fallout from them ending is correspondingly milder. The relationships we have with characters and famous people may not be 'real' in the strictest sense, but the emotions we experience from them *are* very much real. Because our brains are powerful enough to allow us to experience real emotions for individuals that aren't actually part of our world.

Maybe that's the advantage parasocial relationships have over the real kind. The other person doesn't need to be present and interacting for us to feel emotions for them. And for me, this is really reassuring. My father is gone now; I'll never see him again. He's not part of this world anymore.

But the rest of my family and I will carry on loving him anyway. Because, it turns out, our incredible brains are perfectly capable of doing that. It really shows just how powerful the emotional bonds between humans can be.

Sometimes they're so powerful that even death itself may not be enough to sever them.

6

Emotional Technology

I have a red pyjama top in my bedroom cupboard. It's baggy, bobbled, and faded. But it's still comfortable, and that's all that matters with pyjamas, right?

However, I've not worn it in over a year. I'll likely never wear it again. But I can't bring myself to throw it out either. It has too much emotional significance: this pyjama top is what I was wearing when I spoke my last words to my father.

It was a Saturday morning in April 2020, and I'd just been informed by the hospital that Dad, who'd been on a ventilator for over a week, was too far gone. The effects of the virus had proven too severe, and his body had no hope of recovering. It was just a matter of time until he succumbed. That could happen in a few days, or a few minutes, so if I wanted to say goodbye to him, I had to do it now. And that's why I ended up in my kitchen, in my pyjamas, forcing out a tear-choked goodbye, trying to articulate everything my dad had meant to me in nearly forty years and how much I loved him, in a few short sentences, having been given twenty minutes' notice to do so.

And I had to do it over WhatsApp, via a voice call, as a heroic intensive care consultant held the phone up to my dying father's unresponsive ear. For the record, this is absolutely *not* how I wanted it to happen. It just felt so wrong.

But then, would doing it in person have been any better? Presumably, it would have been just as emotionally harrowing, if not more so, but for different reasons. Maybe instead of cheapening or demeaning the experience, saying goodbye over the phone actually served to cushion the emotional blow, and left me able to function and endure.

As it was, I wouldn't have got to say goodbye *at all* without the use of communications technology. The pandemic enforced strict separation and isolation in the hospital wards, so my being physically present was never an option. And, in terms of the wider population, my experience was one of many emotional encounters – first dates, birthdays, hellos, goodbyes – that happened via technology at this time, because there was no other option. Did this diminish them in some way? If so, why?

Everything I've covered so far about emotions is the result of processes in the brain that have evolved over millions of years. However, in the present day we're regularly dealing with things and experiences that didn't exist even a couple of decades ago. What are the ramifications of this? Is interacting with loved ones through screens as emotionally rewarding as doing so in person? Are our social media friends as meaningful as friends in the real world? Where does the divide between real and unreal fall in the digital realm? Given what I'd gone through, and the things I'd experienced, these questions felt very important to me. And I suspected that many others felt the same.

So, I decided to seek out some answers.

Social needier: the emotional impact of social media and related technologies

Remember when I mentioned that hundreds of people would have attended my father's funeral, had that been allowed? I know this, because they watched the livestream of the service.

As the most technologically savvy family member, it fell to me to figure out some way for Dad's extensive network of friends and colleagues to remotely participate in the service. To this end, I created a dedicated group on Facebook for everyone who wanted to pay their respects, and streamed the ceremony to it via my phone, carefully placed at the back of the chapel.

From a purely logical perspective, it made sense to do this via the world's largest social media platform. Everyone who wanted to be there had a Facebook account so was familiar with it, it's free, and it has in-built options for live streaming. Having ready access to such technology would have been unthinkable barely two decades ago. It's amazing, when you think about it.

But, despite all these reasonable points, it still felt *wrong*, to share *my father's funeral*, on *Facebook*! Even writing that sentence is weirdly jarring. I use Facebook to publicise my work, post jokes or memes, and share pictures of my notorious cat. Using it to broadcast my father's funeral service? That was unsettling.

Why, though? Facebook, Twitter, Instagram, Snapchat, TikTok, and more are omnipresent parts of the modern world. I'm a member of several such sites, as is a significant chunk of humankind.[1] Nonetheless, sharing something so profoundly emotional as my father's funeral service via social media felt like a step too far.

However, I'm aware that not everyone feels the same. For instance, my wife and I once used Facebook to invite our friends to her birthday party. This prompted numerous baffling queries about whether we were having marital difficulties. Why? Because *I wasn't included* in the list of Facebook invites. For the record, I didn't get a Facebook invite to my wife's birthday because *we lived in the same house*, what with being married. Why would I need a Facebook invite to my own home? But no. My attending my wife's birthday wasn't validated by Facebook, so, to many, it wasn't happening. Weird, no?

It's far from an unusual occurrence. For many, sharing something on social media seemingly validates it. We all know people who compulsively take photos and share updates about their food, their outfits, their gym or diet progress, the concerts they attend, or the TV they're watching. It's seemingly an essential part of the experience for them.

To clarify, there's nothing objectively wrong about this. But it's a surreal phenomenon: the virtual world carrying *more* significance than the real one, in a sense.

From a neuropsychological perspective, the omnipresence and popularity of social media is an intriguing phenomenon, as is how much influence it currently has. And if I wanted to get to the bottom of my emotional disquiet, I felt I needed to understand the effect social media has on us, and our emotions.

First and foremost, social media expands our ability to socialise. The clue's in the name. Studies have repeatedly shown that positive social interaction stimulates the parts of the brain responsible for reward.[2] Because the digital environment has none of the limitations imposed by physical space or distance, we can interact socially with significantly more people online.

And while our online interactions are free of the physical restrictions of the face-to-face variety, our brains respond emotionally to online exchanges with other people in much the same way as they do to the in-person kind. This isn't dissimilar to what we've seen about parasocial relationships: if our brains constantly form strong emotional attachments to individuals who exist solely in the words in a book, doing the same with a real person's online representation is much less of a leap.

Social media also provides constant novelty, which is similarly known to boost reward activity in the brain.[3] We all have familiar things that we like, that reliably make us happy. But experiencing these things in a *new* way? That makes us happier still. Our favourite band releases a new album, a new instalment in a beloved series of novels is published, catching up with our best friend about what's happened since we last met: all very enjoyable experiences, all of which are things we're already emotionally invested in, which are also infused with novelty.

This combination of familiarity and novelty is a heady cocktail for the brain. Social media is very generous in this area; our feeds provide a seemingly never-ending stream of new updates, posts, links, memes, games, GIFs, etc. from people we like, trust, and admire. Why wouldn't our brains enjoy that?

Then there's status. As we now know, we're subconsciously highly sensitive to how we're perceived by others, and where we stand in the immediate social hierarchy. Consequently, the human brain has developed traits to maintain, or improve, our status.

We covered one example earlier regarding the effects of testosterone, but another particularly interesting instance of how our brains are wired for social status is known as impression management.[4] This is our brain's tendency to (instinctively or consciously) utilise social interactions to give the best possible impression of ourselves. We invariably try to look our best, to agree with those we're interacting with (at least superficially), to hide or deny our flaws or mistakes, or reflexively make excuses for them, and so on. All this is geared towards making people think better of us, which maintains/improves our status.

Human brains were doing this long before social media, but with its arrival our opportunities for status manipulation have increased exponentially. Now we can take hundreds of selfies and only share the most flattering one; spend hours crafting the best possible wording for a post, comment, or tweet; regularly share stuff that makes us look insightful, considerate, generous, caring, or whatever we might desire. And if something gets a bad reaction, we can just delete it, minimising any reputational damage. Basically, social media greatly enhances our ability to present ourselves in the best possible light. It's another thing our brains just lap up when offered.

This instinctive drive to be perceived positively, and all the effort we put into it, means another factor comes into play with real-world social interactions: risk. Our brain's impression management instinct may constantly compel us to *try* to make others like us, but there's no guarantee of success. Other people have complex internal lives of their own, and we can't account for every possible variable of even the most straightforward interaction. Something could easily go wrong, and part of our brain is constantly wary of this. What if I inadvertently say something insulting or upsetting? What if my fly is undone? What if I have spinach in my teeth? What if I accidentally laugh at a tragic anecdote?

It's actually much easier to mess up in a face-to-face inter-action than you might think. For instance, in a real-world conversation, if you take too long to think of a response, it's often seen as a failing. Humans have evolved to interact, to discuss, to gossip.[5] According to some studies, human brains are so inclined to communicate that when two people converse, dedicated parts of their brains 'sync up', effectively becoming two components in one system dedicated to exchanging information,[6] something which plays a part in the process of mimicry, discussed earlier.[7] So, our brains have evolved for a constant stream of information, of dialogue, when we interact. Taking too long to respond, even if it's for perfectly valid and sensible reasons, is therefore quite jarring, and can be seen as someone being uncertain, slow-witted, dishonest, inauthentic,* and more.

This leads into another aspect: real-world interactions are a lot of work, cognitively speaking. Constantly having to process and react to the discussion, in real time, is mentally demanding. On top of that, we're always working out how best to present ourselves.

---

* Because it seems like you're spending time having to come up with some-thing to say, rather than going with what you actually feel.

Add to *that* the constant assessment of the risk of making us look bad,[8] and all told, it's hardly surprising that social interaction is often so exhausting.[9]

Essentially, despite how rewarding they may be, real-world interactions come with a lot of risks and demands. It's like drinking the finest of wines, while crossing a tightrope suspended over a deep gorge. And this is where social media steps in once again, because whoever we're interacting with there isn't physically present, meaning the rules, demands, timings, and expectations are very different, and the odds of inadvertently embarrassing yourself, or upsetting someone, are greatly reduced.

And if you do upset them, they're nowhere near you, so pose no danger. Reduced risk is yet another thing our brain responds very favourably to, particularly when there's no obvious drop in the rewards on offer.[10] Same goes for giving the brain a reduced mental workload; we tend to like and enjoy things more when our brains aren't trying to do too many things at once.[11]

Increased scope for social interactions, enhanced self-presentation, greater safety via reduced risk or embarrassment: social media offers all of this in abundance. But taken together, they also provide a greater sense of control, over how we come across, how we interact, and when and who with. An increased sense of autonomy is *another* thing our brains find very rewarding.[12] Particularly when it's about something very important to us, which social interactions and interpersonal relationships very much are.

Now, all of the things I've listed are typically things our brains react to at the subconscious level. We find social media enjoyable, satisfying, and rewarding (i.e. it makes us happier), but it's in ways that we aren't consciously aware are happening. This is reflected in the way much of our social media use doesn't involve much conscious decision making.

Have you ever stayed up much too late because you were

scrolling through social media into the early hours of the morning, despite knowing that this was unwise and unhelpful? Do you often reflexively check your phone for social media updates? Even in places where it's not necessary or practical, like in a public toilet? Do you regularly but unthinkingly break away from something you *should* be doing in order to check social media?* And how many people have you seen declare that they're taking a break from social media because it's become too consuming, or distracting? This all suggests that much of the allure of social media comes via subconscious, emotion-heavy processes.

However, one thing I've seen repeatedly during all this is that emotion and cognition are highly intertwined. Where emotion goes, cognition often follows, and vice versa. As such, social media being emotionally rewarding means we regularly end up consciously appreciating it, too. Maybe this explains why people often don't feel something is 'valid' unless it's shared online? Social media, via its adept manipulation of our emotional processes, has become such a familiar and engaging aspect of so many lives that it may well have become integrated into more conscious processes and routines. It's like asking your long-term partner before making a decision that affects you both, or even like showering and brushing your teeth before leaving the house: technically you don't *have* to do these things, but not doing them feels several kinds of wrong. The same may apply to not sharing your recent experiences on social media.

It can have more direct effects on our thinking, too. Some studies suggest that using social media can induce a state that neuro-scientists and psychologists identify as 'flow'. It's a tricky concept to pin down (and even trickier to study), and there are numerous theories about it,[13] but here's my understanding of it.

The human brain isn't just one system doing one thing at a time.

---

* I've done it several times while writing this paragraph, I assure you.

It's a ridiculously complex tangle of networks, regions, and processes, all doing multiple things, at the same time. These things make up our minds, our consciousness, but they're by no means working together seamlessly. We've all experienced working on an important task, only to have your attention constantly wandering off. Or trying to sleep, but finding your mind filled with embarrassments or concerns, like unpaid bills, unresolved family conflict, looming deadlines, and so on. Basically, there's a lot going on in our brains at any one time, and most of the time, the different parts of our brains are getting in each other's way, tripping each other up, or even directly competing for superiority.

But occasionally, we end up doing something that stimulates us in just the right way, to make all the argumentative aspects of our brain start working together harmoniously. As a result, not only do we suddenly become very good at what we're doing, we *enjoy* it. We find it really stimulating. You could be a musician performing a brilliant solo, a builder constructing something new and original, a teenager playing a particularly detailed video game: whatever it is, you become totally immersed in the task at hand, and your abilities seem significantly better than normal.

This is cognitive 'flow',[14] also known as being 'in the zone'. It's essentially what happens when our thinking, attention, subconscious, senses, emotions, and more are working in unison, for once. Because there's little or nothing distracting or diverting mental resources to other things, it feels like everything your brain is doing is happening more smoothly, quickly, and easily. It *flows*.

Flow has some aspects in common with the phenomenon of transportation, that thing where we get so emotionally absorbed by a book, film, etc. that our awareness of the real world diminishes. But while transportation is a more passive process (we're just an observer; we can't influence the narrative) flow happens when you're *doing* something, performing a task.

Scientists argue that flow can only be achieved when there's a precise balance between skill and demand. As in, it's when what a task needs you to do, and your ability to do it, line up precisely.[15] If the task is too easy, less of your brain is engaged, so the uninvolved parts carry on doing their own separate thing, preventing a flow state. If the task is too hard, much of your brain is preoccupied with stress, lack of control, self-doubt, uncertainty, and so on, which also prevents a flow state.

But some things sit in the Goldilocks zone of being 'just right', and things flow. It gives us feelings of control, competence, motivation, achievement, positive self-image (because we're vividly aware that we're doing something very well), and more.

Interacting with many people at once is certainly a task, one that our brains are very skilled at and happy to take on. And, as we've seen, social media stimulates many different neurological processes, which is essentially the core requirement of achieving flow. Studies suggest that social media taps into this process, in which case it's no wonder we're constantly compelled to immerse ourselves in it again and again. Flow is something many people spend their whole lives working to achieve.[16]

So, that's *why* social media is so absorbing and compelling, and how it so readily becomes a big part of how we engage with the world. But, as many a scaremongering newspaper article will tell you, social media isn't all good. Far from it.

We can have many more online friends than offline ones, and evidence suggests that the larger their online friend network, the happier and more satisfied people tend to be. However, the same evidence suggests that our closest, most rewarding relationships are still the ones that happen largely face to face. It's not that our social media relationships *aren't* emotionally rewarding; it's more that they can't quite match up to 'the real thing'. The emotional satisfaction offered by online relationships is apparently more about quantity than quality.[17]

This is backed up by studies revealing that excessive social media use, or dependence, is more common in socially anxious people.[18] It makes total sense: if you struggle to form real-world relationships, social media offers the ideal alternative, what with all the reduced risk and greater control. That's presumably why social media's been a boon for many marginalised individuals or groups who, in the physical world, are usually ignored, or worse.[19] On the other hand, if you already have sufficient emotionally rewarding face-to-face relationships, social media doesn't have quite as much to offer you.

This is still focussing on the positives of social media, though. Unfortunately, there's a lot about it that can easily lead to severely negative emotional experiences. Perhaps the most cited example is cyberbullying, the act of bullying or harassing someone via electronic means. Current statistics suggest that the majority of teens and young adults have experienced cyberbullying.[20] Social media is undoubtedly a prominent factor in this, and some aspects of it can make cyberbullying even more harmful.[21]

Cyberbullying might not have the physical component of its real-life counterpart, but this just means its consequences are *all* emotional.* The distress of receiving hurtful messages; the anger at the injustice and your inability to do anything about it; the fear of not knowing when the next message will arrive; who's behind it, and why, and who else is involved, and so on: these can, and do, have detrimental impacts on mental health.[22] This is why experts agree that cyberbullying, just like the physical-world equivalent, is a legitimate form of bullying, just as harmful as the 'traditional' kind.[23] Taking the physical element out of it doesn't reduce the damaging emotional impact, it just cuts out the middleman.

* This isn't even considering how emotions always have an integral physiological component, as we've seen repeatedly. So, scientifically speaking, the distinction between the two isn't clear-cut in any case.

In fact, social media can mean that cyberbullying is *worse* than the traditional kind.[24] Being a member of multiple social media sites means a potential bully has greater access to you than ever. And unlike real-world bullying, cyberbullying can happen to you wherever you are, as long as you're logged in. When bullying is an ever-present threat, it's more stressful again.[25]

But perhaps the most intriguing interaction between bullying and social media is the impact it has on the observer effect. Bullying is typically assumed to involve just two parties: bully and victim. But if others are present to observe the 'exchange', it makes it worse. Just having the loss of status witnessed by others amplifies it considerably, because how others see you is a fundamental part of it. But on top of that, if the observers did nothing about the bullying, it's worse again. Whatever the reason for the bystanders' lack of intervention,* for the victim, it's like they were considered unworthy of help. Because otherwise those watching would have stopped the bullying, right? Thus, the victim feels even worse, emotionally.[26] This is presumably why so many bullies surround themselves with hangers-on, or 'minions'.

The effect of observers in bullying is particularly important for cyberbullying via social media, because it's pretty much impossible to be 'alone'. Unless it's some sort of direct message, all interactions and engagements have the potential to be seen by everyone you and the other party are connected to, and beyond. So, if someone leaves a mean-spirited or flat out abusive comment on your innocuous post, it could be seen by hundreds, thousands even.[27] And if none of those witnesses come to your defence, it may well lead to the same emotional discomfort as that experienced when bystanders don't come to your aid in the real world.

---

* Self-preservation or peer pressure are both valid reasons for this. Doesn't make it any more ethical, though.

Social media is inherently a 'public' space. Everyone who's connected gets to see each other, at all times. That's essentially the point. But that isn't how we interact in the real world; no human has all their friends surrounding them 24/7. This aspect of social media has several unhelpful emotional effects, beyond those involved in bullying.

Foremost amongst these relates to impression management, that subconscious drive to present the best possible image of ourselves. This powerful impulse means that we regularly present an excessively positive portrayal of ourselves, even if it diverges from the underlying reality.[28] To present ourselves as great and worthwhile, it seems we constantly lie to ourselves about how good we are and what we're capable of. Scanning experiments reveal that certain parts of the prefrontal cortex are activated when we tell negative, critical lies about ourselves, but *not* when we tell positive, flattering ones. This suggests that self-aggrandising self-deception is the *default* state of our brains;[29] we only register a change when the lies we tell ourselves cease to be positive ones.

Before anyone gets the wrong idea, this constant self-deception is *good*. Necessary, even. The better we feel about ourselves, the better our mental and emotional wellbeing tends to be.[30] So, having an inaccurate, overly positive self-image is all well and good, because it can subtly motivate us, and guide our face-to-face interactions.

But here's where social media starts causing problems, because the technological freedom and control it offers mean we can present this inaccurate self-image to the (virtual) world. And that's *not* helpful.

Maybe you have an online friend who's constantly posting inspirational memes and 'helpful' advice, even though you know full well that their life isn't going great, they make bad decisions, and are generally miserable. Or someone who constantly posts pictures of themselves at exciting-looking parties in exotic locations, despite being perpetually skint. Social media is awash with people present-ing themselves and their lives more positively than is actually the case.

It's not necessarily bravado (or unfounded arrogance), though. As alluded to earlier, the time we spend on social media can feed directly into our self-perception.[31] This means that the interactions we have online help our brain shape how we see ourselves, just like real-world experiences. So, when someone is exaggerating their positives via Facebook or Instagram, they're not necessarily trying to manipulate other people into liking them. Instead, they might be convincing *themselves* that they're as good as their virtual image suggests.

Unfortunately, everyone else still gets to see this overly positive image. And this can cause issues.

For one, people who make confident claims and assertions are often perceived as more trustworthy than those who seem unsure or uncertain. However, this process is reversed, meaning the confident person is perceived as *less* trustworthy and reliable than the average person, if their claims are subsequently revealed to be deceitful, or inaccurate.[32] This is, presumably, why modern politicians are typic-ally viewed as untrustworthy; their whole existence revolves around confidently making claims that rarely, if ever, turn out to be accurate. Closer to home, it implies that if you regularly exaggerate how great your life is on social media, and someone in your network discovers that things aren't quite as rosy for you as you claim in the real world, this could seriously damage how you're perceived.

This is because the human brain typically doesn't react well to being deceived or manipulated. Self-deception might be standard practice, but someone else deceiving you? That triggers powerful negative emotions. This is why jokes that don't make us laugh can provoke such a hostile reaction;[33] they mean someone assumed they could induce an emotion in us, and then *failed*, so that wannabe emotional manipulator assumed we're simpler than we are. How very dare they!

Research reveals that an excessively positive social media portrayal can be an indicator of issues and insecurities. A particularly notorious example is people constantly posting about how much they love their partner, how they're their 'whole world', incessantly sharing photos of them looking adoringly at each other with flowers and hearts around them, insisting that they're #Blessed, and so on.

You may think this is sweet, or you may think it's nauseating (personally, I'm the latter). In fairness, given the potent impact of romantic love on the brain, some examples of this will undoubtedly be legitimate. However, according to research, such behaviour is often the result of romantic insecurity, meaning those who constantly trumpet their love on social media are often the ones who are *more* anxious about their relationship, not less.[34]

This may seem counterintuitive, but when we consider how fundamental romantic relationships are to our identity, it adds up. If you're experiencing doubts about your relationship, that leads to a lot of emotional discomfort. But then along comes social media, which allows you to present your (virtual) relationship as rock-solid. And if our social media presence really is an important part of how we see ourselves, this makes us feel better.*

The same logic could presumably apply to anyone bigging themselves up to excess on social media, in any other way. It's technically a form of 'fake it till you make it', because much of what our brain does regarding our sense of self is exactly that. But from the

---

* There may also be more cynical motivations behind it. For instance, constantly telling all your mutual acquaintances that your relationship is going great creates social pressure and expectations for your partner, meaning greater consequences for them if they end the relationship. I'm just going to assume that everyone's intentions are decent, though. There's enough negativity out there already.

perspective of those seeing it, it can seem annoying, or false.

This brings us back to the initial point: on social media, almost everything we do or say happens in front of an interconnected crowd. But that's not how most real-world interactions work, and the ancient socialising systems in our brains are confused by this, with harmful consequences.

This process may be at the root of much of the damage done to people's wellbeing by social media. Data suggests a strong link between *subjective* social status and mental health.[35] That 'subjective' is important. Objectively, you may have a very nice life, if you're well educated, live in an affluent area, earn a decent living, have access to all the conveniences of modern life, and so on. But if everyone in your ingroup *also* has these things, only more, or better quality, then *subjectively* you feel very low status. That's an emotionally unhealthy place to be, hence is bad for your mental health.[36]

Social media gives us a much-expanded network of relationships, and much greater, and persistent, exposure to the people in it. And because of how our brains work, most of these people are almost certainly presenting an overly positive image of themselves. So, if someone uses social media often, they can end up constantly seeing friends, acquaintances, and those they admire, all looking great and like their lives are going brilliantly. Such overly flattering representations may be innocuous in isolation, but the cumulative effect is dangerous.

The problem is, for the person using social media, the only flaws or problems they're aware of are *their own*. They get a highly polished perspective of everyone else's life, but a 'warts and all' perspective of theirs. The result is that they, subjectively, feel like the most flawed person in their network, i.e. the lowest-status person of all. Which compromises mental health, and leads to problems.

Some may scoff at this; surely nobody takes other people's social media presence *that* seriously? Sure, much of the time, that'll be the case. But as we saw with parasocial relationships, teen crushes, and the like, the human brain doesn't need much to go on before putting a lot of mental and emotional investment into someone.

In addition, we've already seen that socially anxious people, i.e. those who are very sensitive to how others perceive them, are also the ones who rely most heavily on social media, so would logically be more exposed to an abundance of unrealistically positive portrayals of others. The result of this is, social media could easily make existing feelings of inferiority worse for those who *already* struggle with such things, possibly to the extent of doing genuine harm to their mental health.

This effect is compounded by the fact that socially anxious people are, by definition, less likely to engage with others, either online or in person. This is an issue, because numerous studies show that whether you're a 'passive' or 'active' user of social media changes the effect it has on your wellbeing.[37]

Active users are those who regularly post things, share experiences, reach out, and communicate with others. For such people, social media can actually be *good* for their wellbeing and mental health. But if you're a passive user, one who sits back and merely observes what everyone else is doing, you don't get to experience any of the positive emotional outcomes of connection, interaction, and approval. You just see how good everyone else has it (even if they're not being truthful), which can be very bad for your mental health.[38]

There seems to be a generational divide with this, too. Much of the research into the effects of social media focuses on adolescents and younger people,[39] as the first generation to go through a key stage of neurological and emotional development with social media as a constant presence in their lives. This has, quite reasonably, led to concern about the potential harm social media can do to younger people's mental health, with the default assumption being that it has a negative effect on the wellbeing of teens. That's not what the science says, though. Overall, it seems that, on average, social media has minimal impact on young people's mental health, if it has any at all.[40]

You may find this claim surprising, especially given everything I've said already. However, it isn't the case that social media has *no* effect on young people's mental health. It definitely does. Rather, it's that the effects it has can be both bad and good.[41]

There are many ways for social media to harm the mental health of adolescents and younger people. It increases exposure to abuse, cyberbullying, and social rejection (something the teenage brain is particularly sensitive to[42]). Indeed, by connecting them to so many more people, social media can paradoxically make adolescents feel more isolated, by making them aware of all the things they're not doing, or not invited to.*

Modern technology also allows us to artificially enhance our image, via selfie filters and face-tuning apps. The sharing of these enhanced images, which display standards of physical beauty that are literally unobtainable in reality, often leads to harmful body-image issues, particularly for teens, and especially teenage girls.[43] So, social media clearly *can* harm the mental wellbeing of teens and younger people.

However, social media can also have *positive* impacts on their mental health. It can boost self-esteem, by allowing younger people, who are still figuring out their ultimate identity, to present themselves however they like, with minimal effort and risk. There are countless examples, from decades past, of teens being condemned, scorned, or actively harassed, for daring to look and act differently to the norm.

* The dreaded FOMO: fear of missing out.

The nature of social media means this is less likely to happen.

Relatedly, by connecting them with countless other like-minded individuals, social media can enhance perceived social support, which is crucial for the teen brain. It can also allow safe expression and discussion of otherwise sensitive developmental issues, like sex, by greatly enhancing opportunities for learning from others, and having safe discussions.* And more besides.

The evidence we have seems to suggest that, by and large, the positive and negative impacts of social media on teen mental health often cancel each other out. Again, it's not that social media has no effect on young people's wellbeing. But there's more balance between pros and cons than most assume.

Older generations, on the other hand, respond differently to social media. For instance, one study took post-retirement people who were experiencing isolation and loneliness, and, over several weeks, taught them how to use social media. The idea was that, given how social media connects you to many others and allows you to interact with them, this would reduce feelings of social isolation, and improve wellbeing. That's what happens for younger people, after all.[44] Unfortunately, using social media had barely any impact on older people's wellbeing.[45]

It may be a simple matter of older people, with longer lives and more accumulated experiences, having a more firmly established mental model, i.e. understanding of how the world should work.[46] Things that deviate from or challenge this understanding trigger negative emotional responses,[47] as our brains resist change and defend existing ideas and beliefs.[48] So, if your understanding of the world was acquired when social media, and related technologies,

---

* Some argue that abundant easy access to pornography is causing harm, by giving young people dangerously unrealistic ideas and expectations about sex. It's a valid point, but it's more of an 'internet' problem, rather than a social media one specifically.

weren't around, you'll be more suspicious of them, and thus less inclined to use them. And this is where the problems occur.

I was once speaking at a conference where it was noted that we're currently at a unique point in human society, with a generation of digital immigrants (people who grew up before the internet) raising a generation of digital natives (younger people, for whom the internet has been a constant presence their whole lives). You can't overstate the substantial impact the internet has had, is having, and will continue to have, on all our lives. As a result, children, parents, and even grandparents having drastically different feelings about it and how it's used may be causing even more intergenerational friction and dispute than is normal. Again, it's not the best thing for the mental health of the population.

On a more personal level, it did make me wonder if this was why I'd found the whole Facebook funeral streaming thing so unsettling. Was it just my age? That would be an obvious answer, but it didn't seem quite right. I was still a child when the internet reared its multifaceted head. I guess that makes me a digital immigrant, but one that moved to the digital world at a very young age, so I effectively grew up there. I honestly don't find the digital realm weird or unnerving. I love it, and couldn't function without it.

Then it hit me: however comfortable I was, and am, with the internet and social media, my father felt very differently. He was the classic digital immigrant, and very vocal about his distrust and dislike of Facebook in particular. And yet, upon his passing, I ended up using the very same platform he constantly railed against to share his funeral with all who knew him. I doubt he'd have been happy about that.

I was experiencing too much emotional turmoil at the time to really give it much thought, but I now suspect this is why I found the whole 'streaming the funeral on Facebook' process so emotionally unsettling.

But then, what else could I have done? Doing it may have felt disrespectful to Dad's memory, but *not* doing it, denying his hundreds of friends the opportunity to say goodbye, would surely have felt more so.

The harsh truth is that social media and related technologies are part of our world now. My father is not, not anymore. A lot of people are emotionally uncomfortable with the former, and I'm certainly unhappy about the latter. But none of us can do anything about it.

I guess that's one thing emotions and social media have in common: when they go head to head with the real world, the real world tends to win in the end.

Does not compute: how emotions and technology clash

I previously said I didn't cry the day of my father's funeral until later in the evening, when my family were in bed, and I was alone. I implied that this was when I finally felt safe enough to 'let go', because there was no threat to my ingrained need to appear manly and stoic. That was part of it, sure. However, the trigger for me crying was actually a Zoom call with some friends, who were trying to make me feel better.

Venting my emotions with people who cared about me, but didn't really know my dad so wouldn't be grappling with their own grief, seemed like an ideal move. So, after the funeral I messaged some friends who'd said they were around if I wanted to talk, and a Zoom session was hastily arranged.

Despite it going on for nearly two hours, not *one* person asked me about the funeral, nor how I was feeling about it, and I was too emotionally wrung out to steer the conversation that way myself. Eventually, everyone signed off, leaving me sat there, alone. The feelings of loneliness and dejection, from having my friends deem my own father's funeral unworthy of comment, were what finally set me off crying. Better late than never, I guess.

To be clear, my friends aren't inconsiderate and callous people. I love them no less now than I did beforehand. Because they weren't responsible – as far as I'm concerned, *technology* was.

When I re-read the messages I'd sent to arrange our virtual get-together, they weren't as clear as I'd hoped. I thought I'd written something to the effect of, 'I'm back from the funeral . . . and I need to talk about it with friends'. However, what came across to the others was, '. . . and I want someone to take my mind off it, and talk about literally anything else'. Being so tired and emotionally raw, my communication skills weren't perfect. Unfortunately, the upshot was that my friends didn't mention my father's funeral, because they thought *that's what I wanted*.

Except, I didn't.

Talking with someone experiencing grief is a tricky prospect at the best of times. Doing it over a video link, via low-resolution webcams and screens of varying sizes, makes it harder again. If we'd all been together in person, would the same thing have happened? I doubt it. Face to face, my friends would have had a much clearer idea of how I was feeling, and the intent of my original message would have been much more obvious if it had been conveyed verbally, so included my tone and inflections. But thanks to the pandemic and lockdown, we were stuck with remote, technology-mediated communication, which led to problems.

My point is, for all the good it's done and the power it has, modern technology still struggles with emotions. And because emotions play such an outsized role in human interaction, this can be a significant problem, one that many are keen to see solved.

That technology and emotions don't combine well is not a new observation. As mentioned, robots or machines being unable to experience or understand emotions is a staple of science fiction.

But given how technology can now identify faces in an instant, track miniscule eye movements, recognise and translate languages in real time, map genomes, observe individual atoms . . . why do emotions still cause it such problems?

For a start, communication technology removes much of the emotional information conveyed in face-to-face interactions between two people. For instance, smell and touch are potent elements of emotional communication,[49] but these sensory elements are completely absent in even the most advanced technological communication. Communications technology is much more at home in the audio-visual range. But even here, there are notable gaps. The tilt of someone's head, their posture and stance, the fine subconscious movements that convey tension, anger, happiness, fear. The subtle harmonics in voice and tone. Depending on the medium, technology regularly struggles to detect and/or transmit these things. Even with the most up-to-date software, how much body language can you offer on a laptop Zoom call, when you're only visible from the shoulders up?

Basically, using technology to interact with others invariably means much of the usual emotional information is missing from the exchange. Remember, this doesn't go unnoticed by our (subconscious) brains; they expect this emotional information, and when they don't get it, they get confused. For example, when you're talking on the phone, do you get up and wander around? It's a common phenomenon, but there's no logical reason for it. We *can* move about with our phones, but it's unnecessary. When we're talking to someone in person, we seldom start ambling about the room mid-sentence. So, why do phone calls compel us to do so?

One interesting theory I've heard is this: phone calls lack important nonverbal elements of traditional conversations, like facial expression and body language. So, when we're chatting with another person over the phone, the complex neurological systems that handle human interaction come online, but notice the lack of info usually present during a face-to-face interaction. As a result, we're suddenly compelled to wander about and find it (i.e. the person we're talking to), to fill in the blatant gaps in the exchange.

It's a very intriguing theory, but, despite having been told about it by several different people, I've not found any published research which backs it up. Which is a shame.

Others suggest we pace around while on the phone because the neurological activity allocated to empathy and emotional reactions, having nowhere to go because we can't see the other person and what they're doing/expressing, is sort of 'diverted', and manifests as movement.[50] The brain–body connection gives rise to some interesting things, and physiological responses are a big part of the emotional experience, so this explanation isn't far-fetched. Indeed, studies reveal a strong link between creativity and/or problem solving, and physical movement.[51] And conversing with someone in real time undoubtedly involves a lot of creativity, because it's impossible to script your chats ahead of time.

When we communicate online, there's even more scope for problems. Particularly on social media, where communication happens largely via text, still images, and short videos. While they're an easy, safe, even fun way to converse, they often struggle to incorporate the rich emotional information of a real-world exchange. Our brains have to do a lot of guesswork when figuring out the emotional aspects of such simple messages, and they can easily guess *wrongly*, as my post-funeral Zoom call demonstrated.

Also, emotional reactions triggered by online interactions seemingly aren't as potent as in person.[52] This may be why online relationships typically aren't as emotionally significant as real-world ones. They *can* be – it's now extremely common for romantic relationships to begin online. However, it's rare for them to *remain* online in perpetuity; meeting up in the real world is still a crucial step in any fledgling romance. The online, technological world, for all that it can do, struggles to accommodate meaningful emotional connections. Genuine romance is a tall order without those.

Ultimately, forming meaningful, lasting emotional connections with someone online is not impossible; it's just more of a challenge for our brains, as it's a medium that lacks much of the emotional information they've evolved to expect and utilise.

Also, in the physical world, emotional expression often happens without us knowing. The feelings we have and display, and the empathy we experience in response to those of others, typically occur before our conscious minds get involved. Our emotions and cognition influence each other heavily, but we rarely stop and consciously consider how to express our current emotional state. Nobody actively thinks, 'I am angry about what's just happened, so I am going to pull a relevant facial expression, to ensure that everyone knows my feelings'.

This isn't the case when we communicate via technology. Sure, you regularly see posts on social media that are detailed descriptions of someone's harrowing, or uplifting, emotional experience, or videos or photos of people in floods of tears because they want to share their vulnerability in the moment. This is no bad thing: if you're willing, and able, to be emotionally open and vulnerable in such a controlled and populated environment as social media, more power to you.

However, even in our most emotionally aroused states, we don't write an extensive Facebook post all about our feelings, or film and upload a video where we share them, without *realising* we're doing it.

Unlike our faces, bodies, and various glands, the internet isn't directly connected to our subconscious brain. Therefore, anything we put online has gone through our hands, mouths, and language centres, which are largely controlled by higher-conscious processes. Basically, the stuff we share online can't be 100 per cent reflexive, instinctive, or automatic, because sharing anything online, including emotion, has to be a conscious decision.

Granted, there'll be many benefits to having to think about your emotions before sharing them. That technology gives us greater control over how, and when, we express our emotions, that's one of its strengths. Like everything, though, there are downsides.

As we've seen, including *all* relevant emotional information in an online communication is difficult. This can be problematic enough, but what can make it even more so is that the emotions people communicate online may not even be the *right* emotions.

Emotional and cognitive processes exist in our brains in a sort of dynamic equilibrium, with each influencing (or dominating) the other, depending on the context and situation. But expressing emotions online requires more conscious thought, and increasing the role of cognition in expressing emotions alters this important balance. One consequence of this is that our *online* emotional expression starts diverging from our *in-person* emotions.

Anecdotally, it's said that the most impassioned, confident, or bolshy people online are surprisingly meek, relaxed, or diplomatic in person. There are many explanations for this. They may hide their real feelings in person because other people's physical presence makes emotional honesty a riskier strategy. Alternatively, expressing emotions online, with its greater cognitive component, may lead to overthinking, and the saying of things not in keeping with your default state. Whatever the reason, it's widely acknowledged that people can come across very differently online, compared to in person.

It's hardly a new phenomenon; everyone instinctively behaves differently while we're at work compared to at home, or around friends in a pub. Our powerful brains can sustain multiple expressions of our identity, allowing us to better fit in with certain situations and groups.[53] This applies to our emotions too. For instance, if we're at a comedy night and the comedian tells a crude or gross story, it's (usually) perceived as fun and amusing, so we laugh. But if the same person said the same things out loud on a busy street, we'd find it disturbing. And they'd risk getting punched in the face. Two very different emotional reactions to the same thing, but in different contexts.

The online and real worlds are certainly different enough 'environments' for people to develop different behaviours and reactions in both, which will shape their emotional expression accordingly. Indeed, research reveals that if you assess someone's emotional state via their online output (by analysing the number of emotional terms used, etc.) while simultaneously getting them to record their emotional state in person, the emotional data derived from the two methods can be noticeably different,[54] despite both coming from the same person, during the same time period.

Admittedly, the extent of this effect depends on the individual. If you're someone who prioritises openness and sharing with others, your online and real-world emotions are closely aligned. Conversely, if you value privacy or maintaining a positive image with others, you're more wary of publicly sharing emotions (among other things), so your online and real-world emotions differ considerably. Hence some people are pretty much the same person online as in the flesh, in the emotional sense, while others aren't. In any case, it means that communicating emotions via technology is made even more confusing, because you can't be 100 per cent certain that the emotions being conveyed are an accurate reflection of the individual sharing them.

Speaking of artifice, here's another important point: the internet is not a naturally occurring thing. It's not some digital savannah we humans stumbled upon. Rather, it was deliberately constructed by individuals and organisations. Particularly social media sites:

they're made, owned, maintained, and overseen, by technology companies. But these companies' reasons and methods for doing so are often inscrutable to, or hidden from, those using their platforms. This can cause issues.

In 2014, it was revealed that Facebook, the world's biggest social media platform, had been running experiments on almost one million users, without their knowledge or consent, to the shock and outrage of many. Facebook's defence, that this was covered by the terms and conditions users agree to upon signing up to the service,* didn't satisfy many, least of all because it fell far short of the standards for informed consent required for most experiments.[55]

Why is this relevant? Because the experiment Facebook conducted concerned users' emotions. Specifically, whether they could be manipulated, or controlled.[56] Their results suggested that yes, they could.

Facebook's approach was relatively simple: they manipulated what posts users saw in their feeds. Some saw more emotionally negative posts (bad news stories, sad life events, anger-inducing injustices, etc.) than usual, while others saw more positive ones (cheerful news articles, inspirational memes, etc.). Predictably, those seeing more negative things started posting more emotionally negative things themselves, while those exposed to more positive stuff posted more positive things. So, it was concluded that the emotional content of users' social media feed influences their own emotional state. Therefore, online emotional contagion is a potent force on social media.

However, even leaving aside the dubious ethics of it all, when you know more of the specifics about how emotions work, this conclusion is questionable. Other recent (and more rigorous) studies and analyses suggest a more complex picture of emotional contagion

---

* But which practically nobody ever reads, as they and everyone else knows.

online.[57] While we're undeniably exposed to more people posting emotional content than ever, the lack of subliminal emotional cues available online, and our brain's potential tendency to tire of emotional content and effectively 'tune out', could well prevent this tipping over too easily into emotional contagion.

Also, our social media feeds are produced by the people in our networks. We choose to include these people, so they're mostly individuals we relate to, and feel affinity for. That's basically the point of a social network. So, if such people start posting emotionally provocative things online, saying, 'Look at this injustice, it angers me!', our brains could well react by going, 'That is indeed an injustice that goes against my beliefs and morals, which I share with this person, so it makes me angry too, and I'm also going to express this belief online'. When this happens, we're aware of the source of our emotional reaction. This means it's *not* emotional contagion, because that's when we can't pin the emotions we're experiencing to a specific person or source.

Our brains also really like social harmony. Often without realising, we're prepared to sacrifice a *lot* in the name of not rocking the boat. So, if our social media feeds, populated by a network of friends and like-minded types, start posting mostly unhappy things, we'll likely feel compelled to follow suit. Posting positive emotional things, even if they are an accurate reflection of how we're feeling, is now against the grain of our network, and we don't want to swim against that tide.[58]

Put simply, there are many things happening in our brains that motivate us to change our online emotional output to match what we're being presented with in our feeds, which *aren't* emotional contagion. This may seem like splitting hairs, but it's important, because if a multibillion-dollar organisation that influences a third of humanity is basing its decisions and actions on inaccurate information, that's deeply concerning.

Another concern is this: a key issue with the Facebook study is that it's based on a flawed premise, one alluded to earlier. The emotional states of hundreds of thousands of people, which the results and conclusions were derived from, were determined by analysing the emotional content of their posts. So, the study assumed the subjects' posts on the site (after unknown manipulation of their feeds) were true and accurate reflections of their internal emotional state. However, we now know that this is by no means guaranteed, suggesting that Facebook's conclusions may be even more unreliable.

Why did Facebook do this experiment? What was in it for them? The truth is, determining and manipulating people's emotional state, rapidly and accurately via technology, is the holy grail for many big companies and organisations, particularly those involved in advertising, marketing, and security.

Much research into detecting and influencing emotions via technology comes from the corporate sector, because emotions play a major role in our decision making, including deciding what to buy.[59] Basically, we're more likely to spend money on things we're emotionally invested in.[60] It could be positive emotion (buying an item of clothing after a beloved celebrity wore it), or negative (buying a car that's bigger/better than the one owned by the neighbour you hate), but the result is the same: emotion made you spend money on something. So, if you were a company that had a product to sell, and could detect and influence the emotions of millions, why wouldn't you use this ability to influence them into buying your product, via targeted advertising or direct manipulation?

This isn't a new phenomenon: organisations have been manipulating people's emotions for their own gains for centuries. Governments and ideological news platforms, often working together, regularly use scaremongering and threats of hidden dangers to instil fear in populations, to keep them in line and more easily controlled.[61] And religious figures throughout history delivered 'fire and brimstone' sermons about what awaits people in the afterlife if they don't lead a virtuous existence. That's pretty much the same thing: making people feel fear, to motivate them to stay faithful (and, let's be honest, obedient).

It's not only negative emotions, like fear and anger, that can be used this way. Positive emotions are equally effective. Messages of hope and optimism were at the core of Barack Obama's first successful presidential campaign, and celebrity endorsements – exploiting parasocial relationships by associating a product with someone many people love and admire – are a tried and tested tool of marketing and advertising.[62]

However, modern technology has created a wealth of new opportunities for emotional manipulation of the masses. Previously it was largely a matter of putting an emotionally evocative message out there into the world (via newspapers, TV, billboards, etc.) and hoping enough people see (and are appropriately affected by) it. Now, the internet, social media, smartphones, omnipresent surveillance, and more, mean corporations can interact with countless people on an individual basis, observe their emotional responses, and use this facility to refine and achieve their aims. It's no wonder Facebook studied whether they could emotionally manipulate users. That's very valuable information, financially.

However, while corporations are blatantly keen on detecting and influencing emotions, it's becoming increasingly clear that they don't *understand* them. They don't get or appreciate how emotions really work, or how complex and confusing they invariably are. The Facebook study is a case in point: they assumed that someone's emotional output online is a true and reliable reflection of their internal emotional state, but the science doesn't support this. That's not the only instance of this, though. Far from it.

As well as those concerned with profit-making, security firms/ organisations are similarly keen on finding reliable ways to detect and recognise people's emotions. For example, following the 9/11 bombings, airports have become increasingly focussed on security and the thwarting of terrorist attacks. However, airports also have countless people, from all over the world, coming and going at all hours. This presents a dilemma: how do you increase security measures, which slow people down and keep them out, while simultaneously accommodating an ever-increasing stream of global passengers, who need to be granted access as quickly and easily as possible?

One possible solution is to use technology and software that can scan airport crowds and identify specific facial expressions and behavioural cues to quickly spot anyone who's excessively nervous, or angry, or otherwise suspicious. In 2007, the United States Transport Security Administration did just this, launching the 'Screening of Passengers by Observation Techniques' programme, aka SPOT. Using ninety-four individual screening criteria, it was designed to root out potential terrorists among air passengers by recognising signs of stress, aggression, anxiety, etc. By 2015, after employing nearly 3000 people and costing almost one billion dollars, SPOT had caught zero terrorists,[63] and the programme was mired in complaints and criticism throughout.

The failure of SPOT can be attributed to many things, but consider this: its underlying principles were based on the work of Paul Ekman.[64] We saw, back in Chapter 1, that Ekman's work into how facial expressions are reliable, consistent, and accurate reflections of people's emotional state, has been challenged constantly since its heyday. Of particular import was Professor Feldman Barrett's discovery that, when deprived of context or any hint as to the cause, our ability to determine someone's emotional state from their facial expression is severely compromised.[65]

However, Ekman's original theories were incredibly successful and influential. So much so, they're still assumed by many to be 100 per cent fact, and have shaped the approaches and thinking of many powerful organisations. Many official published studies into the workings of emotions are based around the assumption that facial expressions are an accurate and reliable way to read emotions, which throws doubt on their conclusions.

And so, SPOT, guided by Ekman's theories,* expected people to reliably, quickly, and with minimal exposure, detect and determine the emotional state of complete strangers. The latest science says that such expectations are very unreasonable, as that's just not how our brains do things.

Unfortunately, and worryingly, many powerful bodies that oversee things like justice, law enforcement, and security, still adhere to these outdated and potentially dangerous ideas about emotional expression and recognition, despite the repeated urgings of many in the scientific community to use more evidence-based approaches.[66]

And this is *before* modern technology gets involved. The hope, or expectation, is that specifically designed software could detect concerning emotional expressions much faster, and in greater numbers, than human observers. Unfortunately, if the human brain struggles to recognise a stranger's emotions quickly and accurately via their expression, despite millions of years of evolving complex neurological systems for understanding emotional expression, what chance does recently developed technology have of doing *better*? It's no wonder that developing tech that can actually do this is still proving a challenge.[67]

Despite these obvious limitations, I've lost count of how many

* To clarify, this isn't exactly Ekman's fault. He's adapted and modified his theories in response to new evidence. But the impact of his original work is clearly beyond his control at this point.

products, programmes, and 'exciting new companies' I've seen that claim they can 'monitor your wellbeing', 'determine customer needs and desires', 'shape the user experience', etc. by detecting emotions via facial recognition technology. Even the Chinese government have embraced this, incorporating emotion recognition tech into its surveillance systems.[68] However, acting like such technology exists, and works, doesn't magically make it a reality, and there's abundant data saying it isn't. Not yet. That so many in powerful positions insist otherwise is baffling, and alarming.[69]

Why doesn't it work, though? What's preventing modern technology from mastering emotion, like it's done with almost everything else?

One factor has already been mentioned: context. It may have surprised you to learn, via the whole SPOT debacle, that people are so bad at reading someone else's emotions from their facial expressions, given everything I've covered about how good our brains are at picking up and recognising the emotional states of others. But then, we hardly ever have *just* someone's facial expression to work with. It's undoubtedly very important to the process, but to make sense it needs everything else around it too. It's like how Mona Lisa's smile is probably the most famous in the world, but if da Vinci had painted *just* the smile, on the back of a postcard, it wouldn't be. The individual element is nothing without the whole picture. Our brains feel the same about emotional expression.

For example, if you saw an image of someone's face, wide-eyed and mouth agape, you'd assume they were surprised. If the image then zooms out, revealing that the face belongs to someone who's just received a new car as a gift, you'd assume you were right. Alternatively, if it reveals that it's the face of someone who's just discovered a machete-wielding murderer in their kitchen, you'd probably now rethink, and assume the facial expression is displaying fear. It's the same expression in both cases, but signifies different emotions in different situations, different contexts.

In most situations, there's always some wider context available when deciphering the emotions of others. It's only when artificial constraints are imposed, whether via technological limitations, experimental setups, or practices like the SPOT programme, that our emotional recognition abilities falter.

And even if they didn't, context would *still* be vitally important. The SPOT programme is a perfect example of this, because it was used in airports. Countless people are terrified of flying. Navigating multiple layers of intense security is anxiety-inducing, as is being late for a flight. Having one delayed for countless hours makes us angry. So does being at the mercy of a pompous but untouchable customs agent. The point is, even if it *were* possible to easily recognise when someone's feeling anxious or aggressive, in the context of airports, people experience these emotions for many reasons, most of which are far more common than 'plotting a terrorist attack'.

Once again, this is how tricky it is for humans, with our 'exquisitely sensitive to emotional information' brains. How's a bunch of code on a hard drive or server meant to do better?

There's also another factor at work. Modern technology may be advanced enough to represent or mimic emotions, but we humans often experience negative emotional reactions when it does that. While we can be seriously moved by a heartfelt Facebook post or Twitter thread, a powerful Instagram or TikTok video, here we recognise that the emotion being shared originated from another human, so our brains instinctively fill in any gaps resulting from the medium.

But if emotional information stems from an *artificial* source, i.e. is produced rather than just distributed by technological means, we often feel discomfort and dislike, regardless of what it's trying to convey. For example, many loathe dealing with automated voice systems. Whether it's calling your bank, booking cinema tickets, or listening to announcements about delays while stood on a train station platform, having to interact with a series of recorded messages can be teeth-grindingly frustrating. There are many things underlying this, but one is simply that humans don't like being lied to.[70] It makes us lose trust, become angry at the attempted manipulation, and so on.

Humans are typically very adept at detecting emotional deception.[71] Anyone trying to convince us they're feeling an emotion when they aren't needs to be exceptionally good at it, as our brains are very hard to hoodwink here. That's why we rarely believe anyone who's upset but insists they're 'fine'. It's also another reason why bad acting and canned laughter can be so grating.

So, when a recorded voice tells us, 'I'm sorry for the delay' or, 'Your call is important to us', we aren't mollified, or fooled. How can a recording 'feel sorry' for us? It doesn't even know we exist, let alone empathise with our situation! It's not genuine emotion. Therefore, logically, it's deception, which we instinctively object to.

Despite many great advances in synthetic voices and text-to-speech software, our brains remain keenly aware of the difference between artificial and real voices, and only experience emotional intimacy with the latter.[72] For all its sophistication, technology's efforts to portray genuine emotion to the human brain are still at the level of a twelve-year-old child trying to buy beer from a particularly shrewd barman.

This is why, if you're listening to this as an audiobook, it's been painstakingly read by an actual skilled human, not quickly converted into audio format by a computer programme. While the latter would be faster, and cheaper, nobody would enjoy listening to it, rendering it self-defeating.

Luckily, when the visual element gets involved, it changes matters. It gives the artificial portrayal of emotions more to work with. Specifically, it gives technology *faces* to work with. And while we may need the context to decipher them, faces are still a big part of emotional expression. So much so, our brains actively seek them out, sometimes too enthusiastically, meaning we see faces which aren't really there, like Jesus appearing in our burnt toast. This is the phenomenon of pareidolia, and it's a quirk of our brain's efforts to derive meaning from the modern world.[73]

This process can have surprising emotional consequences. For example, very few people would be emotionally invested in a potato, beyond its use as a foodstuff. But stick a plastic set of eyes, a mouth, and a hat on it, and suddenly it's an iconic toy, beloved by millions of children.

Essentially, we can feel emotions for, empathise with, even create parasocial relationships with, artificial creations, as long as they have features we identify with. That's why comic and cartoon characters are so popular: despite being blatantly artificial formats, the visual nature of their mediums means they can still have many human-like characteristics, via visible faces and bodies. The general rule of thumb seems to be that the more human qualities an artificial creation has, the more emotionally engaging it is.

Next time you watch a cartoon featuring ostensibly human characters, check out how often they blink.* They'll probably do it a lot. Now, cartoon characters, i.e. two-dimensional drawings, don't *need* to blink. Their eyes aren't real, so don't need moistening. Humans need to blink, though, and blink often. Whenever we're talking to someone face to face, they're regularly blinking. And so are we. It's so common, so normal, we don't pay any conscious attention to it. But if someone *doesn't* blink, that's an anomaly, one our brain's subconscious attention processes pick up, and assume something's wrong. Hence, in fiction, not-blinking is often used to denote intensity, or scariness.

---

* *The Simpsons*, one of if not the most popular animated series of all time, is a good one for this.

So, cartoon characters regularly blink. Not because their eyes get dry, but because it makes them more 'human'. We feel more positively emotionally inclined towards them, because they don't set off any emotional alarm bells. As long as artificial characters have enough recognisably human traits, our brains will happily make an emotional connection, despite all their radically different and unrealistic qualities. Indeed, many cartoon characters look nothing like actual humans, but they behave, act, and move *sufficiently* like them, or a familiar sort of cute animal. As long as they tick enough recognisable boxes, it's fine. We're emotionally engaged.

However, certain artificial creations, whether it's primitive CGI, early-stage androids, or unsettlingly realistic dolls or puppets, are *very* similar to humans, but counterintuitively provoke a *negative* emotional response. They strike us as eerie, off-putting.

This is the uncanny valley,[74] the phenomenon where the more human an artificial thing appears the more emotionally appealing it is, until it reaches the very-but-not-quite human stage, when our liking for it nosedives, and rebounds when we perceive it as *actually* human, producing a down-then-up 'valley' pattern on a relevant graph.

Why this happens is unclear, although there are numerous theories, like how it's an evolved instinct to keep us away from corpses. Back in ancient times, corpses were dangerous to be around, given how much infectious bacteria they may contain, and the dangerous predators or scavengers they could attract. And death alters how someone looks; they still appear human, but not quite.* So, we evolved to instinctively recoil from things that are very close to, but not quite, the human norm.

Whatever the reason, the uncanny valley means technological

---

* As an experienced cadaver embalmer, I can confirm this.

portrayals of emotion are even trickier to get right. It *can* be done, but it's not easy. I've already revealed I'm a big Pixar fan, and, via their output, the studio has managed to get millions of people emotionally invested in CGI representations of toys, monsters, cars, rats, and even, in the case of the beloved WALL-E, what is essentially an elaborate box.

But Pixar also clearly know the limits of conveying emotions via technology. None of their human characters have 100 per cent realistic human dimensions, thus avoiding the uncanny valley. And actual humans voice every character, because computer-generated voices remain off-putting.* And while Pixar's efforts tend to succeed, others regularly miss the mark. For every loveable bunch of charming characters found in films like *Toy Story* or *WALL-E*, there's the unsettling array of monstrosities in *Mars Needs Moms*, or gang of glassy-eyed nightmare children in *Polar Express*. Even live action films can be afflicted with distressing digital creations these days, like the twitch-inducing digitally resurrected Peter Cushing in *Rogue One: A Star Wars Story*.

Basically, even multi-billion-dollar companies, with armies of employees and the most cutting-edge equipment, still struggle with detecting and representing emotions via technology. When you look at it that way, it's unsurprising that a colleague's email has the 'wrong tone', or a Facebook post strikes some as a legitimate cry for help and others as risible attention seeking, or that my friends and I got our wires crossed on a post-funeral Zoom call. Using technology to share emotions is a more uncertain process than we realise.

However, if there's one thing technology's always doing, it's

---

* Pixar even used this to their advantage in the aforementioned *WALL-E*. Every machine character in that film is voiced by a human, except the ship's autopilot, the film's antagonist. It has a synthetic voice, making it seem cold and unsympathetic.

advancing. So, technology may currently struggle with recognising and expressing emotions, but that won't necessarily always be the case.

For instance, modern text-based communication now includes the option of hundreds of emojis and emoticons. Much as language purists may loathe them, inserting these little faces, symbols, and figures effectively adds an otherwise absent or hard to convey emotional component. So, understanding the intent or feelings behind someone's words is easier. Especially when you include memes and GIFs. It shows that everyday technology has reached a point where conveying more emotional depth in our communications is practically second nature for many.

What about technology, like computer software, detecting and recognising emotions without the aid of other humans? Things are progressing there too. Complex methods like machine learning and neural networks (where processors are set up to extract and refine information in configurations that mimic the function of biological neurons[75]) are reportedly developing software that *can* recognise emotions online, in ways that *do* take account of the wider context.[76] And they're constantly getting better at it. That's how progress works.

Of course, given all I've said, this may be a *bad* thing. Do we want powerful, unaccountable corporations and organisations to have working technology that can accurately monitor, or even influence, our emotions? After all, they're already trying to arrest us or make us buy things by using methods and technology that *don't* work that well. It's a valid concern.

But it's not all negative. Emotionally accurate and sensitive technology can be a boon for mental health, particularly when it comes to expanded and enhanced therapy. In Chapter 2, I spoke to Dr Chris Blackmore, of Sheffield University, about how his research into integration of emotional qualities into online learning platforms. If anyone's working at the convergence of emotions and technology, it's him. Luckily, he also enlightened me about the development of software algorithms that assess the communication output of patients in therapy (whether it's audio recordings of counselling sessions, discussion forums, social media posts, etc.) This software detects changes in what a patient says which indicate they're considering abandoning therapy, on the verge of relapsing, or about to experience an episode of their condition. If you compare the patient in therapy to someone walking across a frozen lake, these verbal changes are like spreading cracks in the ice, appearing before they plunge into the frigid water. Spotting these warning signs allows the therapist time to guide them onto firmer ground, so to speak.

And this is achieved by the software recognising/quantifying emotionally loaded terms in the patient's communications, like 'depressed', 'scared', 'worried', 'hurt', and so on. If such words occur more often, it indicates negative emotions, linked to their disorder, are building up in the patient's brain, influencing their speech. And it apparently works: such algorithms have been shown to flag up imminent relapse in patients dealing with addiction,[77] psychosis,[78] and more.

As well as recognition, technology effectively conveying/ displaying emotions also has therapeutic implications. Face-to-face talking therapies, like cognitive behavioural therapy (CBT), require much time, effort, and expense, as they involve an extensively trained expert speaking to just one person, for many hours, on a weekly basis. This would be an issue even if mental healthcare weren't chronically underfunded and under resourced globally, which it is.

However, if software could provide such therapy effectively, that would make it considerably cheaper, and far more accessible, for millions, and be a great boon for mental healthcare. So, there's unsurprisingly been a lot of (encouraging) research into developing such virtual therapists.[79]

Of course, for virtual therapists to work reliably and effectively, they'd have to display and detect emotions (in the patient) as well as a human can. That's a big ask. Moreover, face-to-face therapy often works *because* it's another human being doing it: someone the patient can form a trusting emotional bond with, letting them feel safe enough to share their issues, and accept help. Time will tell whether technological alternatives can clear these hurdles.

But even if technology is never able to process emotions as well as humans, it can still play an important role in how we deal with them. Another technological innovation in mental healthcare is avatar therapy.[80] Basically, if you keep hearing voices, aka auditory hallucinations, due to a psychotic disorder, avatar therapy creates a CGI head or face to essentially act as the 'source' of them.

The symptoms of mental health problems are often so disturbing because they're so entwined with our own minds and consciousness, so we don't know where they're 'coming from'. They have no obvious parameters or source, and that's seriously unsettling. Thankfully, technology can now provide something to effectively 'take the blame', giving us a target, a focus, for our emotional distress. Being able to say, 'It's not our fault, it's this bozo on the screen' can make a world of difference to our wellbeing.

Virtual reality, VR, is also increasingly useful when dealing with mental symptoms. For example, if sufferers of PTSD encounter anything that reminds them of their trauma, it triggers an extreme, and debilitating, emotional reaction. Luckily, therapists can now help patients process such triggers in a healthier way by experiencing them via safe and controllable VR. The outcomes of such approaches are encouraging, so far.[81]

Maybe current technology isn't that great at detecting,

communicating, or displaying emotions in convincing ways, but there's still much it can offer. Technology can provide us with an outlet for our emotions that isn't another person, so won't be upset, sensitive, or have their own feelings to bring into the exchange. It's weird to think of technology's *lack* of emotional abilities being a positive emotional thing, but it makes sense.

After all, 'technology' doesn't just mean the modern sort, where everyone has several powerful gadgets, often kept right in their pocket. The stone axe was once cutting-edge technology (pun intended). Since then, we've had pen and ink, printing presses, cassette recorders, and more. They're all examples of technology, and each granted us an additional way of expressing our feelings. The importance of this, and how it shaped us, really can't be over-stated. As my own experience of grief during a period of global isolation demonstrated all too well.

Of course, there's a downside to all this. Because, even if technology can effectively communicate emotional information, what guarantee do we have that this information is correct, or valid? The answer is, 'none whatsoever'. As anyone who's ever been online will undoubtedly recognise, the 'news' you encounter may well be 'fake'.

It turned out that emotion is at the heart of this phenomenon, one I felt compelled to look into because, as well as the many drastic consequences it's had for our whole society, it ended up making the most painful period of my life even worse.

Fake news, real views: how emotion and technology undermine reality

Grief is a very emotional experience. This is hardly a revelation at this point. But there's a difference between knowing something in theory, and it actually happening to you. That's how I finally understood that grief isn't just prolonged sadness, one omnipresent emotional state, but made up of many different ones. Sadness is in there, obviously, but so is fear. And the many forms of emotional pain. Regret. Guilt. Shame. Some of these don't even make logical sense, but that's never stopped emotions before. It's a real mixed bag, a big, heavy bag, of distressing feelings.

But one emotional aspect of grief that *did* surprise me was the anger. As covered earlier, a loved one's death is a profound, powerful loss, one you can do nothing about. It *always* feels horrendously unfair, because there are zero circumstances where it will ever feel justified. Perceived unfairness and loss of control reliably make humans angry, and grief provides them in spades.

I'd wager, though, that the anger I experienced while grieving was worse than usual, because the thing that caused me the most anger, following my father's death, was probably having complete strangers insisting that it didn't happen, or that it was irrelevant.

Grief typically doesn't involve having your pain and turmoil openly mocked or dismissed. But, in 2020, that was the reality I, and countless others, faced. Because we'd lost loved ones to COVID-19. And, in defiance of all evidence and the very laws of reason, a large and distressingly vocal number of people maintained that the virus was actually harmless. Or didn't exist. Or 'only killed people who are already sick and unhealthy', because apparently having less than 100 per cent perfect health means your life is worthless.* Despite my occupation, I'm struggling to articulate how it feels to be reeling from the most emotionally painful experience of your life while armies of strangers insist it didn't happen! 'Infuriating' just seems insufficient.

Predictably, the vast bulk of these dubious claims were found online, usually via social media. Some might say, 'Just don't go online, then'. Unfortunately, during a lockdown, that was the only

* I was also told lockdowns were unnecessary because COVID-19 would only kill 1 per cent of people. In the UK alone, that's around 700,000 lives, a higher death toll than that inflicted by World War II.

way of engaging with people, something I desperately needed, given what I was going through. Also, it wasn't just anonymous trolls, lurking in the darkest corners of the web. These enraging claims often came from prominent media figures, politicians, even world leaders! In our media-saturated, interconnected world, how is anyone meant to avoid that?*

If I couldn't stop it, and lacking any better options, I resolved to figure out *why* it was happening. That should at least allow me a sense of control over the situation. So, how could so many mature adults become convinced that the terrible things that had happened (and blatantly were still happening) to me and countless others, were made up, exaggerated, or some kind of conspiracy? To invoke a bleakly common modern term: why were so many convinced that the pandemic was 'fake news'?

Here's the thing: the human brain likes acquiring new facts and information about the world, and the people in it. Much of what we've covered feeds right into this. Our brains make us innately curious, they crave novelty, they're constantly scanning for potential dangers or advantages, they find uncertainty stressful, they're forever coming up with simulations and hypotheticals as to what could happen to us, and so on.

Learning about what's currently happening in the world around us facilitates all these things, and is used to develop our understanding of how the world works, to guide our decisions and actions, to shape our beliefs, attitudes, actions, decisions, our very thinking. Overall, the information and facts that our brains take in determining our understanding and perception of the world. That may seem like a needlessly obvious point to make, but it's a crucial one.

The question is, *how* do we acquire information? Where do the

---

* Also, if you want me *really* angry, please continue lecturing me on how I 'should' grieve.

facts that our understanding is built upon come from? Primarily, like most species, we acquire information about our world via our senses. That's what they're for. That plant is green, these berries taste good, it hurts when this predator bites me: these are immediate, tangible facts that our senses supply to our brains.

However, human brains are capable of much more than that, and because we evolved to be intensely social beings, we also rely heavily on other people for information. Many parts of our brains are dedicated to extrapolating information from other humans (e.g. the empathy networks), often just by observing them. Interacting with them adds a whole other dimension. It means we acquire information indirectly, in the abstract. Someone else tells us, 'Don't go down to that river, there's a hungry tiger there', and so we follow their advice and survive, because we acquired information from someone else, without having to risk our own skin and learn it directly.

Is it any wonder that our brains are so receptive to communication and information-sharing with others? It kept our species alive, and shaped us accordingly, as we've seen. Some theories even suggest the evolution of language, and cognitively advanced human communication, were driven by a fundamental need for gossip.[82] And what's gossip, if not the sharing of new information with others?

But then, humans came up with technology. Among many other things, this allowed us to store and share information more reliably, effectively, and robustly than having to rely solely on messy, ever-shifting human memory. The development of writing was a particularly important milestone here, one that shaped the world as we know it.[83] Whether on stone slabs, clay tablets, or animal skins, writing allowed people to record specific thoughts, ideas, observations, instructions, etc. in an unchanging format that could be readily shared with others, over ever-greater distances thanks to advances in transportation.

And because populations covering far greater areas than tribal villages could now communicate and share the same information, it drastically expanded the size of the groups humans could feasibly 'belong' to.[84] Instead of tribes, we now had communities, villages, townships . . . nations. The major religions exist as a result of this too, because they're pretty much all based around a holy book, or sacred text. It's a lot easier to spread the word of God(s) when it's written down, in a readable format.

This wasn't all good, obviously. A lot of these larger communities ended up being empires, and the history of them (and many religions) include considerable bloodshed. Better information sharing is all well and good, but too often that information was, 'Those people over there disagree with us. We must kill them.'

Nonetheless, for better or worse, when human society gave rise to information-sharing technology, it had an undeniable and dramatic effect on human society in turn. The information we take in determines how we think and act, so having ever greater access to it directly shaped the development of our civilisation, and ourselves. It also accelerated advancement, because when hard-won information is written down and accessible, people don't have to painstakingly rediscover and relearn it all the time.

Jumping ahead to modern times, after a few millennia of cultural upheaval, progress, and technological development, sharing up-to-date information about what's going on in the world with millions of people at once is now a common occurrence. It's a process and industry in its own right, known to many as 'the news'. For most of the twentieth century, when it came to where people got their news, most relied on either newspapers or broadcast media, namely TV and radio. These sources were widely regarded as the most reliable and credible,[85] something which remains largely true today. This meant that, whether they intended to or not, these platforms wield tremendous power and influence.

If the information that our brains absorb directly affects our understanding of the world, then it logically follows that those who control and supply that information can determine what we end up thinking and believing. And research has revealed that this does indeed happen. For example, a 2014 study asked people how common certain types of cancer were. It found that their answers weren't based on actual medical statistics, but on how certain cancers were represented in the news and media. People tended to overestimate brain tumours (comparatively rare, although referenced often in popular dramas), but underestimate bladder cancer (common, but rarely features in the media).[86] And when everyone in a population is getting their information from a few select sources, it means the news can determine the priorities of entire countries.[87]

However, even the most powerful wide-reaching news provider can't say whatever they want, whenever they want, however they want. Technology may have vastly expanded our civilisation's ability to share information, but there are still limits and restrictions on our ability to do this, and most of these are imposed by the human brain. Our brain may like to constantly acquire new information, but it regularly has to work hard to do so.

This is particularly true of pure, abstract information. Raw data, mathematical values and equations, context-free times and dates and definitions: we can understand and retain all these things, but it's not something our brains do easily. It takes time and neurological effort to process this stuff. For example, gossiping is easy, but *studying* is hard. Both involve acquiring new information, but the former involves emotional stimulation and motivation, while the latter is more about taking in abstract information, divorced from context or any particularly stimulating qualities, so only our most complex cognitive processes are involved. For our brains, it's like writing a formal letter with a fine paintbrush: definitely doable, but it takes longer and requires more focus, because that's not strictly what it's meant for.

Mentally processing abstract information involves a surprising number of interconnected neurocognitive regions,[88] and consumes a lot of the brain's resources,[89] hence studying can often feel so draining. There's also the fact that working memory – the brain's facility for manipulating and managing abstract facts and information (sort of like the central processor in a computer) – has a surprisingly small capacity for doing so. It can only hold around four 'things'* at once.[90] If you've ever struggled to remember a whole address or telephone number in one go, that's why.

As a result of all this, bombarding the human brain with information and expecting it to take it all in at once is like trying to push a birthday cake through a drinking straw. Can't be done. But you *can* break it into small pieces and push them through gradually. It takes longer, sure, but you'll get there in the end. It just requires patience and perseverance. Much like studying.

Luckily, our brains are used to this. After all, every second we're conscious, our senses are relaying more information to the brain than they could ever make use of. Accordingly, our brains have developed numerous ways of dealing with this, like the subconscious systems that regularly divert our attention to anything amid the sensory noise that seems important or useful.[91] Similarly, when we're confronted by more news and information than we can cope with, our brains prioritise, and divert attention and resources to that which it considers to be most important.

But how does our brain determine which information is most important? Ideally, it would go through all the available information and work out, sensibly and logically, which is most relevant

---

* What counts as a 'thing' varies between situations, because that's the brain for you.

or urgent. However, that would require our brains to take in all the information beforehand to properly assess it, which is like trying to open a locked box with the key inside it. So, our brains must use something else to determine which information to prioritise. And that something else is usually, and predictably, emotions.

After all, memories with strong emotional components are more effectively processed than those without.[92] We learn better from someone, or something, that we've an emotional connection with.[93] Our sense of smell is particularly stimulating and evocative largely due to its direct connections to the brain's emotion-processing regions,[94] and more. Given all this, emotion being a key factor in what information we focus on and retain is hardly surprising.

TV news and newspapers have clearly long been aware of this, and have incorporated emotional aspects into how news is presented. While it would presumably be much easier to broadcast basic factual text descriptions of important events and occurrences, TV news is still read out by newsreaders, because our brains are far more inclined to take in information when it's supplied by another person, someone we can emotionally engage with. Similarly, newspaper front pages invariably feature big eye-grabbing headlines with emotionally charged words,[95] like 'Scandal', 'Shock', 'Horror', 'Fury', 'Glee', etc., and most newspapers include contributions from relevant individuals, or are even just someone's personal opinion on certain matters. The emotional, human element matters.

Modern news sources also rely heavily on evocative imagery and sound, like dramatic music, detailed photographs, eye-catching graphics, and more. You might consider this excessive, or distracting, but research reveals that accompanying informational statements with emotionally evocative images enhances how believable people find them.[96] So, if you've ever wondered why social media often seems awash with supposedly inspirational quotes and messages printed over beautiful nature scenes or mountain vistas, now you know.

It's all well and good saying news should only include 'the facts', and many platforms insist they do just that, but, as far as our brain's concerned, that would be like going to a restaurant and being given a plate of raw chicken and soil-covered vegetables. That may *technically* be what we ordered, and we could feasibly eat it, but it'd be an unenjoyable struggle. Incorporating emotional qualities into factual information is the equivalent of preparing and cooking raw ingredients, allowing our brains to better consume and digest them. It's an interesting system, worked out over decades, centuries even.

And then, at the end of the twentieth century, technology changed the world, again, by bringing about the 'digital revolution'.[97] Among the many consequences of this, it meant that most people now have access to some form of personal computer and all the functions it provides, including internet access.

Countless things, good and bad, have been attributed to the arrival of the internet. Many of them have been discussed in this chapter already. But among the most profound and impactful is the effect it's had on the average person's ability to access news and share information. In the age of the internet, rather than having to rely on a limited range of news programmes and news-papers being supplied once a day or every few hours, everyone now has all the latest news and information the world has to offer, twenty-four hours a day, accessible at the touch of a button or flick of a screen.

Ironically, back when computers and the internet were more intriguing possibilities than everyday reality, many were looking forward to this exact scenario. If everyone in the world had access to all the factual information they'd ever need at all times, they reasoned, ignorance would quickly become a thing of the past. However, rather than an era of pure understanding and logic, it's now increasingly common online to encounter people who genuinely think the world is flat.[98] The optimistic predictions about the effects of abundant information for everyone overlooked one crucial aspect: the limitations of the human brain.

The internet has given us both far more information than we could ever take in, and considerably more control over what of this abundant information we choose to consume. When presented with significantly more information than ever, our brains must work even harder to figure out what to prioritise and focus on, and to do so, it can end up relying on emotions more than ever. This isn't ideal, for many reasons.

For one, if you only expose yourself to information that is emotionally pleasing or reassuring, your understanding of the world is going to end up skewed and flawed. Because a lot of what goes on in the world isn't reassuring, and your feelings on the matter are immaterial.

Let's not write off humankind just yet, though. Sure, it's statistically inevitable that some will only focus on news that reassures and validates what they already think. However, studies reveal that this is not nearly as common as many fear.[99] There's a lot going on in your typical brain, meaning other complex factors come into play, which prevent everyone immersing themselves in a self-gratifying echo chamber.

Simple human curiosity is one. Very few of us are content to only hear what we already think or know at all times. We're intrigued by the novel, the exciting, even the controversial and taboo,[100] so are regularly motivated to seek these things out, which counteracts the instinctive desire to only engage with information that supports what we already think.

Also, our brains have a negativity bias,[101] whereby anything that elicits a negative emotional response tends to have more of an impact on the workings of our brain than those that trigger positive emotions. This, unsurprisingly, influences the sort of news and information we're interested in as well.

If it feels to you as though modern news is always so bleak and depressing, this is because news sources are influenced by what people want to hear about, what they're emotionally invested in.* And research reveals that people are indeed typically more interested in negative news than positive, *even if we're convinced the opposite is true.*[102] People may claim they're sick of negative news, and genuinely mean it, but they remain drawn to it despite themselves.

This has been demonstrated outside of the lab too, thanks to news publications that opted to report only positive stories, and promptly lost two-thirds of their audience.[103] So, while it may make for a somewhat bleaker existence in general, this negativity bias at least keeps many of us from focussing solely on news and information that sustains a comforting delusion.

Here's another factor: with so much choice available, who do we trust to provide us with information about the world? As I've said, for many years it was predominately broadcast media and newspapers. Preparing a whole newspaper, or several TV news programmes, and providing them to millions every day requires substantial resources and manpower. So, only those able to supply such things could get into the news business, meaning it was largely the reserve of powerful groups or organisations, like businesses, corporations, or government.

But thanks to modern technology, that's no longer the case. Now, anyone with a laptop or smartphone and internet connection can produce information and put it online, with minimum effort. And, largely thanks to social media, everyone has their own public

---

* Consider how much news coverage is dedicated to sports, and celebrity antics. These things rarely have any direct impact on the average person's life. But countless people are still heavily emotionally invested in them, which makes them newsworthy.

platform, where they can share whatever with their wider network, which can easily number in the thousands, or more.

We've seen the many ways in which this can be both a good and a bad thing. But one particularly important aspect is something I alluded to just now: for most of our history, the human brain got much of its information from other humans. And by and large, we still *prefer* to get our information from other people, often at an instinctive level.

Studies and experiments have repeatedly demonstrated that what those around us think, believe, and do, directly influences what *we* think, believe, and do.[104] The human brain is just that social. This means we're strongly inclined to conform, to agree and go along with those around us, those we identify with. Indeed, recent research reveals that, even if they consciously want to, it's genuinely very difficult for an individual to resist the compulsion to conform.[105]

Pre-internet, when the news and information we received about the world came via TV and newspapers, it meant everyone in a population was receiving roughly the same information from only a few sources, so there was a smaller range of likely beliefs and worldviews. In addition, there were regulations and checks and balances to stop newspapers and broadcasters saying whatever they wanted, to suit their own ends, or those of their owners. Oversight bodies, laws against libel and slander, powerful competitors: all worked to keep the output of news platforms 'acceptable'.*

There's also the fact that they needed to maintain credibility and goodwill with their potential viewers/readers, and surveys show that the most important thing people look for in a news source is accuracy.[106] The upshot of all this is that 'official' news platforms, for

---

* Your view on how effective or necessary these things are may vary, but they exist, and that's the main thing here.

all their many flaws, have long had to put a reasonable amount of work into making sure the information they share is valid, accurate, and reasonable. This meant that, among other things, dubious, unverifiable claims were rarely being publicised. After all, if those in charge of a platform had a lot of news to share and only one TV bulletin or front page to play with, they wouldn't squander it by handing it over to someone with ludicrous ideas and an axe to grind, who'll likely get them in trouble. And if someone like that did slip through the net, it didn't go well for them.

Consider the case of David Icke, a well-known footballer, then broadcaster, in the 1970s and 80s in the UK. In the 1990s, Icke began claiming to be the son of God and insisting that a cabal of shapeshifting space lizards controlled the world. This, predictably, led to widespread ridicule and condemnation,[107] despite Icke enjoying a very privileged position compared to the average person, as someone who got regular access to the mass media.

Call me censorious and 'close-minded' if you like,* but I reckon this is helpful. If the information in the news shapes people's understanding of the world, what's not in the news . . . can't. Therefore, it's good that those with extreme and downright dangerous views about other races/sexes/religions, conspiracy theorists, doomsday predicters, etc. didn't have their views shared and amplified, and therefore *validated*, by credible news platforms.

Sure, in a sufficiently large and complex society, fact-free, fringe, or extreme views and beliefs pop up regularly. But it would have been an uphill struggle to maintain and spread such worldviews when most people got their information from mainstream news sources. Therefore, those who did subscribe to such beliefs were far less likely to encounter others who shared them.

Imagine someone sitting with friends in a 1980s bar and revealing

* You won't be the first.

that they honestly think the Queen is really a shapeshifting space lizard vampire. It would probably lead to years of derision and mockery. In other words, they'd have to choose between their unconventional ideas and beliefs and social acceptance. And the latter often wins out,[108] because our brain's all too often willing to jettison information it believes to be correct and valid, if adhering to it means social rejection. So, all told, a population's reliance on established, mainstream media sources for news and information resulted in a more hostile environment for unrealistic, unscientific, and unpleasant worldviews.

For the most part, the checks and balances that prevented mainstream news platforms from saying whatever they liked still apply today. However, there's considerably less regulation and restriction regarding what people can say online, or on social media, and what there is isn't sufficient, according to many.[109] This means any individual with information they deem important, no matter how farcical and unrealistic, can beam it around the planet in seconds. As a result, the amount of unhelpful or inaccurate information people can end up just being exposed to is going to skyrocket.

This is bad, because our brains aren't too choosy, and misinformation can be just as influential as actual information. Remember the study about how people estimate the occurrence of certain types of cancer based on how often they appear in the media rather than medical data? It's a clear demonstration that misinformation (albeit unintended in this case) can still shape people's perception and understanding of the world.

Sure, the data reveals that most people want and expect accuracy in their news sources, but that can be misleading. Something can be objectively accurate, in that it's an actual fact that's supported by all available evidence and data, but few people have the time, resources, and expertise required to actually check such things. For most people, whether something's judged as accurate is more a matter of whether it conforms to what they already know about how things work. But what people 'know' about how the world works is almost entirely dictated by the knowledge they've prioritised and retained, and, now more than ever, this can vary wildly from person to person.

Say you've ended up genuinely believing that the political party that runs your country regularly engages in cannibalistic satanic rituals in hot dog restaurants. If you see two official news reports, one which says that this is the case, and one which says this is nonsense, you're more likely to consider the one that confirms what you already 'know' to be the most accurate. And we can end up genuinely believing outlandish and far-fetched things because misinformation can be just as influential as actual information, as long as we *don't know* it's misinformation.

If anything, because misinformation isn't constrained by things like 'proof' and 'evidence' and the time and effort these things require, it has even more scope to shape the things people think and believe in, no matter how far from objective reality they can end up. Sadly, there are many properties of the modern internet that, presumably without meaning to, have allowed this to happen. To the extent that misinformation, particularly about important things like health, is currently one of the major problems facing modern society.[110]

There's a common phrase among sceptic and rationalist types who work to combat unscientific and superstitious claims: the plural of anecdote is not data. It means that, even if many people tell you something is true, that doesn't make it so. For instance, at a certain point in history, the majority of people living would have confidently stated that the sun goes around the Earth. But it didn't, doesn't, and never has. Objective facts and truth are what they are, regardless of how many people insist otherwise.

Unfortunately, while 'the plural of anecdote is not data' is a valid stance regarding the real world and objective reality, it's a different story on the neurological level. As far as our brains are concerned, if enough people tell us something, we're more likely to accept it as a fact. And the greater our emotional connection to them, the more we're likely to trust them.[III] It's just how we're wired.

It goes back to the difference between gossip and study, and is presumably why newsreaders and celebrity endorsements are so common in the modern media landscape. We spent most of our evolutionary history getting our information from other people, people we're looking at and listening to, so we've developed to find them emotionally stimulating, meaning our brains are more receptive to information obtained from them.

This can be a helpful thing. It's been shown that, when it comes to changing our minds or opinions, we're far more likely to do that if the information required to do so is supplied by another person, or people, that we're emotionally engaged by, rather than if we're just supplied the information itself.[112] This means our preference to listen to other people can be used to combat misinformation and harmful beliefs.

However, this works in either direction, and most people end up being influenced by misinformation and harmful beliefs thanks to their interactions with other people. And if there's one thing the internet has expanded beyond all recognition, it's our interactions with others.

It doesn't help that the internet, and social media in particular, has also blurred the lines between what's a 'credible' source of information, and what's not. Previously, it was easy to discern between an official news source and an amateur one. Put a major newspaper beside a self-published pamphlet and nobody's going to mix them up. Similarly, the polished appearance of official TV news broadcasts could never be replicated by someone with a home video camera in a

basement. Sometimes, credibility can be determined by p
values.[113]

Now, though, countless people get their news and inform
primarily online. So, established news platforms have had to start
channelling their output through Twitter, Facebook, YouTube,
and so on. The online realm can be a great leveller, so telling the
difference between official news sources and random individuals'
input is increasingly difficult. Amateur blogs can look just like professional articles. A major newspaper's Facebook posts can appear
in your feed and look exactly like those of your mother's friend
from work who has some very concerning views about immigrants.
A twenty-four-hour news channel's YouTube videos sit alongside
those by a guy who has a decent grasp of editing software and very
'intriguing' ideas about the illuminati.

There has been a lot of research into the relative credibility of
print, broadcast, and online news sources thanks to this.[114] And
while official, established news providers are still considered very
credible, many studies have shown that, online, people often find
their friends' and close connections' output to be as, or more,
trustworthy,[115] or that a news source is deemed more credible if it
comes from, or conforms to the general output of, your network of
friends.[116] The emotional connections we have with others strongly
influence what information we find credible and are willing to
accept, often without us knowing it's happening.

I believe that last thing gets to the heart of the matter. Perhaps
the most profound consequence of the internet and social media is
that, now, if you have an idea or belief or even suspicion, no matter
how ludicrous it is, you're virtually guaranteed to find information
that backs it up, and many lack the ability to critically assess the
validity of this information, meaning they're more likely to accept
it at face value.

But even more important than finding (mis)information that

supports what you think, is that you're virtually guaranteed to find someone who agrees with you. Often, many someones. And having information validated by others, particularly those we're emotionally connected to, is often the most important factor in whether it's retained and trusted. Once a community consensus is achieved, people will instinctively work hard to preserve and enhance it.[117]

It doesn't matter how farcical it is, or how overwhelming the evidence against it, if other people agree with and support your theory or claim, our brains will see that as validation, as 'confirmation'. We'll be emotionally rewarded for sharing it, not rejected, which will make us even more convinced we're on to something.[118] And so, thanks to how the internet works regarding information sharing and interpersonal connections, an objectively false belief can be, subjectively, validated, nurtured, and encouraged.

You might wonder why, given the internet and social media connects us to vast numbers of people, we aren't similarly influenced by people with opposing or alternative viewpoints. If we're exposed to everyone's opinions all the time, and we struggle to discern between that and actual facts and information, why isn't what we think in a constant state of flux? Good question.

Firstly, we *aren't* exposed to everyone's views all the time. Modern technology means we have a surprising amount of control over who we're exposed to, who we engage with. That's a big part of the appeal. It's like turning up at a huge party where everyone's talking at once: you don't have to engage with everyone, just those you know; the rest become background noise.

However, even more importantly, once information has been accepted into our brains and started influencing our thinking and understanding, our brains are alarmingly reluctant to change or reject it. Appropriately enough, given how our emotions are a key factor in what information we take in to begin with, our established understanding is defended by emotions, via several methods.

There's confirmation bias,[119] where we avoid or ignore information that challenges what we already think. If that doesn't work, there's motivated reasoning,[120] where individuals process information in ways that lead to desired outcomes and decisions, rather than adhering to what the information actually says, objectively. If both these fail to keep out challenging facts, there's belief perseverance,[121] where people will still maintain an existing belief or conclusion even when presented with solid contradictory evidence or information. Existing beliefs can even get *stronger* in response, hence this phenomenon is sometimes also known as 'the backfire effect'.

After all, absorbing, processing, and retaining abstract information is vitally important, but often hard work for our brains, so anything that challenges the information they've laboriously accumulated is technically a threat, as it could potentially undo all that hard work and throw our understanding of the world into disarray. That's why being presented with information that contradicts what we already think and believe often leads to a rapid negative emotional reaction, involving the experience of stress and psychological discomfort, known as cognitive dissonance.[122]

To stop this dissonance, we can either alter our emotional response to it and accept we're wrong, or think about it more critically or cynically than we would more neutral things, allowing us to figure out reasons to reject it, and preserve our existing views and beliefs. It speaks to the powerfully fundamental nature of emotions that it's often a lot easier to change what you think than it is to change what you feel. We've seen this a few times already. People's enthusiasm for bad news – even if they insist they don't like it – and their tendency to conform against their own wishes are both examples of our conscious thoughts and our subconscious emotional drives being at odds with each other. And more often than not, it's our emotions that win out – even if those emotional reactions are based on misinformation.

This has not gone unnoticed. Disturbingly, the misinformation found online doesn't just come from people who are passionate but misinformed; much of it is put there deliberately, by those who want to actively deceive or mislead people. There are so many motivators for this: political power, influence, money, status, ideology, attention, approval, self-esteem, and more. Manipulating people through online deception can lead to all these. And, at present, there is seemingly little consequence for doing it. So, if you lack the scruples to stop you doing it, why wouldn't you peddle misinformation?

Yet again, the exploitation of emotion is apparent even here. If you look at any examples of misinformation, or misleading claims, they're never anything nice, are they? It's always, 'you're being lied to', 'powerful people are trying to kill you', 'all your most alarming suspicions are correct', 'that group you don't like are secret child murderers', and stuff like that. This takes full advantage of the brain's negativity bias, making attention more likely, and makes the recipients of the misinformation emotionally aroused, which, as we know, makes them more likely to retain the misinformation. Studies have even confirmed that those who rely on emotion more than reason in their general thinking are more susceptible to fake news.[123]

Unfortunately, because it's geared towards engagement and likes and clicks and shares, the data suggests that social media is actively set up in ways that encourage emotion and outrage,[124] inevitably making the situation even worse.

Because of all this, many experts and concerned groups now work round the clock to combat misinformation, to put out corrections and counterpoints, to fact-check and flag up false information. It's an uphill struggle, though, given how our emotions put up several layers of defences to keep anything from undermining what we think and believe. And so, wrong, inaccurate beliefs endure, and grow, encouraged by like-minded electronically interconnected communities and the emotional validation they readily provide.

Some might think I'm overstating the problem. After all, it's largely restricted to the online world, and didn't I say just recently that the real world was more emotionally stimulating, and therefore more influential, than the virtual one? True, but I didn't say that the virtual, technological world had *no* influence or effect on us. Sometimes, it can indeed be *more* affecting. Recent studies revealed that overexposure to news coverage of traumatic or disastrous events can have more severe effects on us *than being physically present* at the disaster in question![125]

Maybe this is because, as awful as directly experiencing a major disaster is, once it's over, it's over. That's not true for news coverage, online speculation and reaction, etc. This regularly turns minutes-long incidents into hours, days, even weeks of emotionally potent information, giving it far more opportunity to infuse into the concerns and fears within the recipient's brain.

Perhaps that's where technology and the virtual world actually have the edge. The real world has obvious, tangible limits and restrictions. The internet, social media, etc. do not.

And so, put it all together, what have we got? A modern world where we've more information than our brains can possibly handle, so end up more reliant on emotions when choosing what is important. Where the lines between credible sources and spurious gossip or baseless conjecture are increasingly blurred, and often presented side by side as if they're the same thing. Where any idea or belief, no matter how far-fetched or unrealistic, can quickly end up with supporting 'evidence' and a community that endorses it, which is catnip for our ever-social brains. And where the setup of the online world, coupled with a load of bad actors out to achieve their own nefarious ends (including the politicians and powerful figures who control our society[126]), work to keep us as emotionally stimulated, and thus as susceptible to and accepting of misinformation as possible.

Given all that, it's not really a surprise that so many people believed the pandemic wasn't real. It may be a comforting delusion, but it's one that technology and emotion are quite happy to indulge and validate.

Can we do anything about this? Some suggest the best approach is to separate emotion from the process altogether, and approach everything as rationally and logically as possible, while only deferring to credible, evidence-based sources. I myself was once an enthusiastic proponent of such an approach, and an active member of communities that championed this.

I've learned a lot since then, though, particularly over the course of writing this book. Now, I can't help feeling that this approach is flawed, because it seriously misunderstands, or underestimates, just how vital and fundamental a role our emotions play, in everything we think and do.

Yes, our emotions cause many problems; that's undeniable. But you know what? Bones break. Cells mutate into cancers. Our skin sunburns. Our eyes become misshapen. These things happen all the time. That's life. But nobody ever suggests removing our skeletons, peeling off our skin, or killing off our cells, because we still need these things to function, to exist. I now appreciate that the same applies to emotions.

That's why I think any efforts to ignore/suppress/reject our emotions are doomed to fail. And that's not just a suspicion; you see it often online, with the lofty types who insist that they're beholden only to reason and logic, yet somehow keep having angry arguments with anyone who disagrees with them (something neither reasonable nor logical).

Then there are self-described sceptics, rationalists, intellectuals, etc. who spend years encouraging people to only use scientific evidence and credible sources when making decisions. Don't get me wrong, it's a noble aim, but, increasingly often, such people find something they're particularly passionate about (i.e. emotionally stimulated by), be it gender issues, political ideology, free speech/censorship concerns, or whatever. And suddenly, their noble principles vanish, and anything and everything that supports what they think about their particular matter, no matter how shoddy, controversial, or flimsy, is valid evidence all of a sudden.

It's a scientifically recognised phenomenon. Indeed, a recent study demonstrated that people were far more likely to trust information and news stories that were emotionally stimulating, even if they come from a source *known* to be untrustworthy.[127]

If anything, it's another demonstration of how our emotions can't really be separated from our more cognitive, cerebral processes. Indeed, ask yourself why we humans tend to think rationally, logically, at all? You may say it's because we like to be right, to be correct, to work things out and provide certainty in an uncertain world. It's reassuring. It makes us feel better.

That's all well and good, but what it means is that, ultimately, our brains use reason, logic, and rational thought, because doing so is *emotionally rewarding*.[128] Far from being an impediment, logic and reason *depend* on emotion. They couldn't exist without it. So, any attempt to suppress or eliminate emotions from our thinking is both hugely counterproductive and ultimately destined to be unsuccessful.

So, what's my solution? Emotions can be so problematic and irrational and lead us to believe so many ludicrous, harmful things, but are also a crucial, fundamental, and undeniable aspect of everything that makes us what we are. How do we square that circle?

If you ask me, we don't. For now, we just need to acknowledge it. The true workings of emotions have eluded humankind's finest minds for thousands of years, and will continue to do so for a long time yet. By contrast, I'm just one neuroscientist, sat in a shed on the outskirts of Cardiff, trying to get to grips with my own emotions following the tragic loss of my father, and writing it all down in case anyone else is interested.

I will say, though, that numerous studies suggest that the more aware we are of our emotions and what they do, the more we're able to mitigate and control the effects and influence they have on us.[129] I've certainly found that's been the case for me.

For example, I now understand why some people, when faced with the grim reality of a pandemic, would end up believing it's not real, or has been unleashed deliberately for nefarious reasons. The modern tech-saturated world makes it very easy to do this, and it must be emotionally comforting, for many reasons.

But then I pop up, talking about the emotional trauma I've gone through after the pandemic killed my father. This threatens to rip away such people's emotional comfort blanket, engaging their brain's defences, leading them to conclude I must be lying, or part of a conspiracy, or anything else which means I can be ignored. They're not actively trying to make my grief worse, as much as trying to avoid emotional discomfort of their own.

I can't say I like it. I certainly don't agree with it. But I can at least say I now *understand* how and why it can happen. And that honestly does make me feel better. It's something, at least, because I don't know what could change the minds of such people.

Apart from the death of one of their own beloved family members from COVID, of course. I categorically don't wish for that, though. The last thing that I'd want is for anyone else, no matter who they are or what they've done, to experience the same thing that I've gone through. I don't believe it would make me feel any better if they did. I confess I may have thought that once, not too long ago. But if nothing else, I can say, with certainty, I'm no longer that emotionally ignorant.

Here's hoping I've helped you say the same.

Conclusion

As I write these words, my late father is watching over me.

I don't mean in the spiritual sense.* I mean I've two framed photos of my father on the shelf behind me in my home office.

They're not especially significant or meaningful photos, of major events, or celebrations, or anything like that. Just random pics of my dad and me. Indeed, such was my lack of concern for the photos after taking them that they sat for years on my hard drive, gathering the binary code equivalent of dust. I'd occasionally glance at them while looking for something else, but that's it.

Then, Dad died, in the circumstances he did. And now, those photographs have become far more profound and important to me, and take pride of place in my home office.

I tell you this because it's a perfect demonstration of something explored in this book: that potent emotional experiences have the power to alter and change your memories, and your feelings regarding anything connected to those memories, long after they were formed.

Indeed, now that I think about it, this very book is an example of that. It was originally intended to be a fun, light-hearted affair. Much of it was already written in that vein. But then I went through a deeply emotional experience, and now it's far more profound and personal than I ever planned. Or expected.

I'd argue, strongly, that it's all the better for it. And correspondingly, so am I. Experiencing this journey, and writing it down for you, has drastically changed my understanding of emotions. Which

* Although that might be the case. This is beyond my remit, though.

I expected. However, it became increasingly clear that, right before my eyes, my emotions have profoundly *changed me*, too. And that I didn't expect.

It was probably inevitable, though. Because if there's one common element to everything I've learned about emotions in this book, it's that emotions are all about *change*.

I'd felt I was particularly ignorant about emotions and how they work, but discovered that pretty much everyone else is uncertain about such things. Because our definitions, parameters, and general understanding of emotions, are constantly changing. And have been for millennia.

I'd always believed that emotions were purely abstract occurrences within our brains, but in fact every emotion we feel leads to a physical change within us, be it neurological, physiological, even chemical.

I learned that emotions first arose because the earliest brains needed to know how to react to detectable changes in the environment. And that, far from being a superfluous evolutionary relic, emotions changed us, shaped the way we look, the things we perceive, the memories we form, over millions of years.

Like many, I'd also believed there were specific emotions, for specific occasions. Instead, the emotions we experience change and morph over time, and change wildly from person to person.

And, thanks to how they're infused throughout our brains, emotions regularly change us on the small scale, on the individual level. They can change what we think about things, even when it makes no objective sense, like making us enjoy pain, or recoil at the sight of a loved one.

Because of this, emotion is invariably the conduit via which so many things affect and change us, like music, stories, animals, babies, colours, relationships, and almost everything else we may feasibly encounter.

Emotions can even change our understanding of reality, in good and bad ways. They can compel us to envisage more hopeful outcomes for all of existence, or distort our perception of reality so severely that we reject the evidence of our senses, and attack those who are already suffering.

And whoever you are – whatever your gender identity, whatever your age – your emotions, and their expression, are affected by the world around you, and what it expects of you, just as much as anything happening within your own brain.

And then there's me. How have I changed as a result of this whole saga?

Beyond all the knowledge I've acquired, and the many deeply emotional memories I've accumulated, it's made me realise that I shouldn't resist or reject my emotions, as so many like me are wont to do.

Yes, of course the loss of someone you love, particularly if it's in harrowing circumstances, results in deeply unpleasant, often seemingly unbearable emotions. But I now appreciate that such emotions are, in so many ways, the psychological equivalent of soreness and inflammation following injury and infection. Such things are actually the result of your body responding to the problem, not the problem itself. Likewise, the intense negative emotions that hit us following a tragedy are our brain's way of *dealing* with the experience.

This realisation helped me considerably. To pull the curtain back slightly, I'm writing this conclusion several months after submitting the first draft of the book, and a lot has happened since. Things that had the potential to upset me greatly, to stoke up outrage and despair, both on the global and personal levels. And while they reliably did just that, nothing so far has managed to overwhelm me. I've bent, but not broken.

Indeed, many people have commented on how surprisingly

calm I've managed to remain, given all I've gone through. And when they do, I attribute it to this book, the writing of it, and the things I experienced to get it done. It taught me not to fight or suppress my emotional reactions, but to let them happen, accept them, and see where they take me.

Granted, I'm a neuroscientist, so maybe it's easier for me to say this. But then, I'd argue that being a neuroscientist going through grief like I did was like being an experienced mechanic trapped in a speeding car, with no brakes, on the motorway. Even if I did know what the problem was, and how to fix it, *right then and there*, such knowledge was of little use; my only choice was to cling tightly to the wheel and swerve round obstacles, until things slowed down enough.

I haven't crashed yet. Hopefully, I never will. My driving has improved; I feel more in control. But then, there's still a lot of road to go.

Do I still miss my father? Yes.

Does his absence still cause me emotional distress? Yes.

Do I expect to feel this way to some degree for the rest of my life? Yes.

Is there anything wrong with that?

No.

I can't guarantee I'll never experience such emotional turmoil again. But if I do, at least I'll be less emotionally ignorant about it. Or, if I am, I'll be willing to accept that, and work with my emotions, rather than try to resist or control them. Because in so many ways, they're a vital part of me. And the same is true for everyone else.

If you take anything from this book, let it be that.

Acknowledgements

If my experience is anything to go by, there are many people responsible for bringing a book into existence, even if it is just the one name on the cover.

That's even truer for this book. I'd thought that my previous books were a challenge to write, but now it feels like they were the literary equivalent of some light sparring, before stepping into the ring to battle with a bareknuckle boxing champion.

In all honesty, this book could easily have been ruined, derailed, or completely abandoned, if it weren't for the help and support of so many people, during what proved to be the most difficult time of my life. The least I can do is acknowledge how invaluable they were.

My wife Vanita, and my children Millen and Kavita. The circumstances meant we had no choice but to stick together, but there's nobody else I would have chosen to be with anyway.

My agent Chris Wellbelove, whose legendary calm and unflappability I didn't manage to break, despite my (inadvertent) best efforts.

My editor Fred Baty, my publisher Laura Hassan, and everyone else at Faber, for their boundless patience with my chaotic submissions and laughable adherence to deadlines. Granted, there was the whole 'Major bereavement during the pandemic-induced shutdown of global civilisation' aspect to consider, but even then I was really pushing it.

Dan Thomas and John Rain, great friends who kept me going at the darkest time by getting me to talk drunken guff about shoddy films every Tuesday night.

Richard, Carys, Katie, Gina, Chris, Brent, and everyone who contributed to this book, in ways great and small.

Whoopi and Tom. Because having the endorsement of a literal household name can be surprisingly motivating, even at your lowest ebb.

And finally, Pickle, just for being as fun as he is, even if he'll never know how I appreciated it. But then, even if he did, he'd probably not care in the slightest. That's cats for you.

References

1: The Emotional Basics

1   Firth-Godbehere, R., *A Human History of Emotion* (Fourth Estate, 2022).
2   Russell, B., *History of Western Philosophy: Collectors Edition* (Routledge, 2013).
3   Graver, M., *Stoicism and Emotion* (University of Chicago Press, 2008).
4   Annas, J.E., *Hellenistic Philosophy of Mind*, Vol. 8 (University of California Press, 1994).
5   Algra, K.A., *The Cambridge History of Hellenistic Philosophy* (Cambridge University Press, 1999).
6   Seddon, K., *Epictetus' Handbook and the Tablet of Cebes: Guides to Stoic Living* (Routledge, 2006).
7   Montgomery, R.W., 'The ancient origins of cognitive therapy: the reemergence of Stoicism', *Journal of Cognitive Psychotherapy*, 1993, 7(1): p. 5.
8   Ambrose, S., *On the Duties of the Clergy* (Aeterna Press, 1896).
9   Gaca, K.L., 'Early Stoic Eros: the sexual ethics of Zeno and Chrysippus and their evaluation of the Greek erotic tradition', *Apeiron*, 2000, 33(3): pp. 207–238.
10 Dixon, T., *From Passions to Emotions: The Creation of a Secular Psychological Category* (Cambridge University Press, 2003).
11 Bain, A., *The Emotions and the Will* (John W. Parker and Son, 1859).
12 Wilkins, R.H. and I.A. Brody, 'Bell's palsy and Bell's phenomenon', *Archives of Neurology*, 1969, 21(6): pp. 661–662.
13 Darwin, C. and P. Prodger, *The Expression of the Emotions in Man and Animals* (Oxford University Press, 1998).
14 McCosh, J., *The Emotions* (C. Scribner's Sons, 1880).
15 Dixon, T., *Thomas Brown: Selected Philosophical Writings*, Vol. 9 (Andrews UK Limited, 2012).
16 Izard, C.E., 'The many meanings/aspects of emotion: definitions, functions, activation, and regulation', *Emotion Review*, 2010, 2(4): pp. 363–370.
17 Murube, J., 'Basal, reflex, and psycho-emotional tears', *The Ocular Surface*, 2009, 7(2): pp. 60–66.
18 Smith, J.A., 'The epidemiology of dry eye disease', *Acta Ophthalmologica Scandinavica*, 2007, 85.

19   Dartt, D.A. and M.D.P. Willcox, 'Complexity of the tear film: importance in homeostasis and dysfunction during disease', *Experimental Eye Research*, 2013, 117: pp. 1–3.

20   Vingerhoets, A., *Why Only Humans Weep: Unravelling the Mysteries of Tears* (Oxford University Press, 2013).

21   Frey II, W.H., et al., 'Effect of stimulus on the chemical composition of human tears', *American Journal of Ophthalmology*, 1981, 92(4): pp. 559–567.

22   Bellieni, C., 'Meaning and importance of weeping', *New Ideas in Psychology*, 2017, 47: pp. 72–76.

23   Gelstein, S., et al., 'Human tears contain a chemosignal', *Science*, 2011, 331(6014): pp. 226–230.

24   Rubin, D., et al., 'Second-hand stress: inhalation of stress sweat enhances neural response to neutral faces', *Social Cognitive and Affective Neuroscience*, 2012, 7(2): pp. 208–212.

25   Garbay, B., et al., 'Myelin synthesis in the peripheral nervous system', *Progress in Neurobiology*, 2000, 61(3): pp. 267–304.

26   Heinbockel, T., 'Introductory chapter: organization and function of sensory nervous systems', in *Sensory Nervous System* (InTech, 2018), p. 1.

27   Elefteriou, F., 'Impact of the autonomic nervous system on the skeleton', *Physiological Reviews*, 2018, 98(3): pp. 1083–1112.

28   Jansen, A.S., et al., 'Central command neurons of the sympathetic nervous system: basis of the fight-or-flight response', *Science*, 1995, 270(5236): pp. 644–646.

29   VanPatten, S. and Y. Al-Abed, 'The challenges of modulating the "rest and digest" system: acetylcholine receptors as drug targets', *Drug Discovery Today*, 2017, 22(1): pp. 97–104.

30   Jansen, et al., 'Central command neurons'.

31   Elmquist, J.K., 'Hypothalamic pathways underlying the endocrine, autonomic, and behavioral effects of leptin', *International Journal of Obesity*, 2001, 25(S5): pp. S78-S82.

32   Kreibig, S.D., 'Autonomic nervous system activity in emotion: a review', *Biological Psychology*, 2010, 84(3): pp. 394–421.

33   Bushman, B.J., et al., 'Low glucose relates to greater aggression in married couples', *Proceedings of the National Academy of Sciences*, 2014, 111(17): p. 6254.

34   Mergenthaler, P., et al., 'Sugar for the brain: the role of glucose in physiological and pathological brain function', *Trends in Neurosciences*, 2013, 36(10): pp. 587–597.

35 Olson, B., D.L. Marks, and A.J. Grossberg, 'Diverging metabolic programmes and behaviours during states of starvation, protein malnutrition, and cachexia', *Journal of Cachexia, Sarcopenia and Muscle*, 2020, 11(6): pp. 1429–1446.

36 Kahil, M.E., G.R. McIlhaney, and P.H. Jordan Jr, 'Effect of enteric hormones on insulin secretion', *Metabolism*, 1970, 19(1): pp. 50–57.

37 Gershon, M.D., 'The enteric nervous system: a second brain', *Hospital Practice*, 1999, 34(7): pp. 31–52.

38 Sender, R., S. Fuchs, and R. Milo, 'Revised estimates for the number of human and bacteria cells in the body', *PLOS Biology*, 2016, 14(8): p. e1002533.

39 Mayer, E.A., 'Gut feelings: the emerging biology of gut-brain communication', Nature reviews. *Neuroscience*, 2011, 12(8): pp. 453–466.

40 Evrensel, A. and M.E. Ceylan, 'The gut-brain axis: the missing link in depression', *Clinical Psychopharmacology and Neuroscience*, 2015, 13(3): p. 239.

41 Ali, S.A., T. Begum, and F. Reza, 'Hormonal influences on cognitive function', *The Malaysian Journal of Medical Sciences: MJMS*, 2018, 25(4): pp. 31–41.

42 Schachter, S.C. and C.B. Saper, 'Vagus nerve stimulation', *Epilepsia*, 1998, 39(7): pp. 677–686.

43 Porges, S.W., J.A. Doussard-Roosevelt, and A.K. Maiti, 'Vagal tone and the physiological regulation of emotion', *Monographs of the Society for Research in Child Development*, 1994, 59(2–3): pp. 167–186.

44 Breit, S., et al., 'Vagus nerve as modulator of the brain-gut axis in psychiatric and inflammatory disorders', *Frontiers in Psychiatry*, 2018, 9: p. 44.

45 Groves, D.A. and V.J. Brown, 'Vagal nerve stimulation: a review of its applications and potential mechanisms that mediate its clinical effects', *Neuroscience & Biobehavioral Reviews*, 2005, 29(3): pp. 493–500.

46 Ondicova, K., J. Pecenak, and B. Mravec, 'The role of the vagus nerve in depression', *Neuroendocrinology Letters*, 2010, 31(5): p. 602.

47 Bechara, A. and A.R. Damasio, 'The somatic marker hypothesis: A neural theory of economic decision', *Games and Economic Behavior*, 2005, 52(2): pp. 336–372.

48 Wardle, M.C., et al., 'Iowa Gambling Task performance and emotional distress interact to predict risky sexual behavior in individuals with dual substance and HIV diagnoses', *Journal of Clinical and Experimental Neuropsychology*, 2010, 32(10): pp. 1110–1121.

49 Dunn, B.D., T. Dalgleish, and A.D. Lawrence, 'The somatic marker hypothesis: a critical evaluation'. *Neuroscience & Biobehavioral Reviews*, 2006, 30(2): pp. 239–271.

50 Damasio, A.R., 'The somatic marker hypothesis and the possible functions of the prefrontal cortex', *Philosophical Transactions of the Royal Society of London*, Series B: Biological Sciences, 1996, 351(1346): pp. 1413–1420.

51 Dunn et al., 'The somatic marker hypothesis'.

52 Lomas, T., *The Positive Lexicography*, 2019. Available from: https://www.drtimlomas.com/lexicography.

53 McCarthy, G., et al., 'Face-specific processing in the human fusiform gyrus', *Journal of Cognitive Neuroscience*, 1997, 9(5): pp. 605–610.

54 Gunnery, S.D. and M.A. Ruben, 'Perceptions of Duchenne and non-Duchenne smiles: a meta-analysis', *Cognition and Emotion*, 2016, 30(3): pp. 501–515.

55 Kleinke, C.L., 'Gaze and eye contact: a research review', *Psychological Bulletin*, 1986, 100(1): p. 78.

56 Liu, J., et al., 'Seeing Jesus in toast: neural and behavioral correlates of face pareidolia', *Cortex*, 2014, 53: pp. 60–77.

57 Darwin and Prodger, *The Expression of the Emotions*.

58 Ekman, P., 'Biological and cultural contributions to body and facial movement', in *The Anthropology of the Body*, J. Blacking (ed.) (Academic Press, 1977), pp. 34–84.

59 Ekman, 'Biological and cultural contributions'.

60 Ekman, P. and W.V. Friesen, 'Constants across cultures in the face and emotion', *Journal of Personality and Social Psychology*, 1971, 17(2): p. 124.

61 Sorenson, E.R., et al., 'Socio-ecological change among the Fore of New Guinea [and comments and replies]', *Current Anthropology*, 1972, 13(3/4): pp. 349–383.

62 Davis, M., 'The mammalian startle response', in *Neural Mechanisms of Startle Behavior*, R.C. Eaton (ed.) (Springer, 1984), pp. 287–351.

63 Ekman, P., W.V. Friesen, and R.C. Simons, 'Is the startle reaction an emotion?', *Journal of Personality and Social Psychology*, 1985, 49(5): p. 1416.

64 Jack, R.E., O.G. Garrod, and P.G. Schyns, 'Dynamic facial expressions of emotion transmit an evolving hierarchy of signals over time', *Current Biology*, 2014, 24(2): pp. 187–192.

65 Ekman, P., 'An argument for basic emotions', *Cognition & Emotion*, 1992, 6(3–4): pp. 169–200.

66 Beck, J., 'Hard feelings: science's struggle to define emotions', *The Atlantic*, 24 February 2015.

67   Jack, R.E., et al., 'Facial expressions of emotion are not culturally universal', *Proceedings of the National Academy of Sciences*, 2012, 109(19): pp. 7241–7244.

68   Barrett, L.F., *How Emotions Are Made: The Secret Life of the Brain* (Houghton Mifflin Harcourt, 2017).

69   Gendron, M., et al., 'Perceptions of emotion from facial expressions are not culturally universal: evidence from a remote culture', *Emotion*, 2014, 14(2): p. 251.

70   Bowmaker, J., 'Trichromatic colour vision: why only three receptor channels?' *Trends in Neurosciences*, 1983, 6: pp. 41–43.

71   Hemmer, P. and M. Steyvers, 'A Bayesian account of reconstructive memory', *Topics in Cognitive Science*, 2009, 1(1): pp. 189–202.

72   Güntürkün, O. and S. Ocklenburg, 'Ontogenesis of lateralization', *Neuron*, 2017, 94(2): pp. 249–263.

73   Luders, E., et al., 'Positive correlations between corpus callosum thickness and intelligence', *NeuroImage*, 2007, 37(4): pp. 1457–1464.

74   Frost, J.A., et al., 'Language processing is strongly left lateralized in both sexes: evidence from functional MRI', *Brain*, 1999, 122(2): pp. 199–208.

75   Mento, G., et al., 'Functional hemispheric asymmetries in humans: electrophysiological evidence from preterm infants', *European Journal of Neuroscience*, 2010, 31(3): pp. 565–574.

76   Christie, J., et al., 'Global versus local processing: seeing the left side of the forest and the right side of the trees', *Frontiers in Human Neuroscience*, 2012, 6: p. 28.

77   Perry, R., et al., 'Hemispheric dominance for emotions, empathy and social behaviour: evidence from right and left handers with frontotemporal dementia', *Neurocase*, 2001, 7(2): pp. 145–160.

78   Davidson, R.J., 'Hemispheric asymmetry and emotion', *Approaches to Emotion*, 1984, 2: pp. 39–57.

79   Murphy, F.C., I. Nimmo-Smith, and A.D. Lawrence, 'Functional neuroanatomy of emotions: a meta-analysis', *Cognitive, Affective, & Behavioral Neuroscience*, 2003, 3(3): pp. 207–233.

80   Isaacson, R., *The Limbic System* (Springer Science & Business Media, 2013).

81   MacLean, P.D., *The Triune Brain in Evolution: Role in Paleocerebral Functions* (Springer Science & Business Media, 1990).

82   Nieuwenhuys, R., 'The neocortex', *Anatomy and Embryology*, 1994, 190(4): pp. 307–337.

83   Isaacson, *The Limbic System*.

84   MacLean, P.D., 'The limbic system (visceral brain) and emotional behavior', *AMA Archives of Neurology & Psychiatry*, 1955, 73(2): pp. 130–134.

85 Isaacson, *The Limbic System*.

86 Iturria-Medina, Y., et al., 'Brain hemispheric structural efficiency and interconnectivity rightward asymmetry in human and nonhuman primates', *Cerebral Cortex*, 2011, 21(1): pp. 56–67.

87 Morgane, P.J., J.R. Galler, and D.J. Mokler, 'A review of systems and networks of the limbic forebrain/limbic midbrain', *Progress in Neurobiology*, 2005, 75(2): pp. 143–160.

88 Roseman, I.J. and C.A. Smith, 'Appraisal theory: overview, assumptions, varieties, controversies', in *Appraisal Processes in Emotion: Theory, Methods, Research*, K. Scherer, A. Schorr, and T. Johnstone (eds) (Oxford University Press, 2001), pp. 3–19.

89 Murphy et al., 'Functional neuroanatomy of emotions'.

90 Davidson, R.J., 'Well-being and affective style: neural substrates and biobehavioural correlates', *Philosophical Transactions of the Royal Society of London*, Series B: Biological Sciences, 2004, 359(1449): pp. 1395–1411.

91 Murphy et al., 'Functional neuroanatomy of emotions'.

92 Panksepp, J., T. Fuchs, and P. Iacobucci, 'The basic neuroscience of emotional experiences in mammals: the case of subcortical FEAR circuitry and implications for clinical anxiety', *Applied Animal Behaviour Science*, 2011, 129(1): pp. 1–17.

93 Richardson, M.P., B.A. Strange, and R.J. Dolan, 'Encoding of emotional memories depends on amygdala and hippocampus and their interactions', *Nature Neuroscience*, 2004, 7: p. 278.

94 Adolphs, R., 'What does the amygdala contribute to social cognition?' *Annals of the New York Academy of Sciences*, 2010, 1191(1): pp. 42–61.

95 Zald, D.H., 'The human amygdala and the emotional evaluation of sensory stimuli', *Brain Research Reviews*, 2003, 41(1): pp. 88–123.

96 Pessoa, L., 'Emotion and cognition and the amygdala: from "what is it?" to "what's to be done?"', *Neuropsychologia*, 2010, 48(12): pp. 3416–3429.

97 Davidson, R.J., et al., 'Approach-withdrawal and cerebral asymmetry: emotional expression and brain physiology: I', *Journal of Personality and Social Psychology*, 1990, 58(2): p. 330.

98 Adolphs, R., et al., 'Cortical systems for the recognition of emotion in facial expressions', *Journal of Neuroscience*, 1996, 16(23): pp. 7678–7687.

99 Posse, S., et al., 'Enhancement of temporal resolution and BOLD sensitivity in real-time fMRI using multi-slab echo-volumar imaging', *NeuroImage*, 2012, 61(1): pp. 115–130.

2: Emotion Versus Thinking

1   Smith, B., 'Depression and motivation', *Phenomenology and the Cognitive Sciences*, 2013, 12(4): pp. 615–635.

2   Wayner, M.J. and R.J. Carey, 'Basic drives', *Annual Review of Psychology*, 1973, 24(1): pp. 53–80.

3   Brown, R.G. and G. Pluck, 'Negative symptoms: the "pathology" of motivation and goal-directed behaviour', *Trends in Neurosciences*, 2000, 23(9): pp. 412–417.

4   Higgins, E.T., 'Value from hedonic experience and engagement', *Psychological Review*, 2006, 113(3): p. 439.

5   Macefield, V.G., C. James, and L.A. Henderson, 'Identification of sites of sympathetic outflow at rest and during emotional arousal: concurrent recordings of sympathetic nerve activity and fMRI of the brain', *International Journal of Psychophysiology*, 2013, 89(3): pp. 451–459.

6   Lang, P.J. and M. Davis, 'Emotion, motivation, and the brain: reflex foundations in animal and human research', *Progress in Brain Research*, 2006, 156: pp. 3–29.

7   Valenstein, E.S., V.C. Cox, and J.W. Kakolewski, 'Reexamination of the role of the hypothalamus in motivation', *Psychological Review*, 1970, 77(1): pp. 16–31.

8   Swanson, L.W., 'Cerebral hemisphere regulation of motivated behavior', *Brain Research*, 2000, 886(1–2): pp. 113–164.

9   Risold, P., R. Thompson, and L. Swanson, 'The structural organization of connections between hypothalamus and cerebral cortex', *Brain Research Reviews*, 1997, 24(2–3): pp. 197–254.

10 Risold et al., 'The structural organization of connections'.

11 Swanson, 'Cerebral hemisphere regulation'.

12 Diamond, A., 'Executive functions', *Annual Review of Psychology*, 2013, 64: pp. 135–168.

13 Arulpragasam, A.R., et al., 'Corticoinsular circuits encode subjective value expectation and violation for effortful goal-directed behavior', *Proceedings of the National Academy of Sciences*, 2018, 115(22): pp. E5233-E5242.

14 Berridge, K.C., 'Food reward: brain substrates of wanting and liking', *Neuroscience & Biobehavioral Reviews*, 1996, 20(1): pp. 1–25.

15 Blanchard, D.C., et al., 'Risk assessment as an evolved threat detection and analysis process', *Neuroscience & Biobehavioral Reviews*, 2011, 35(4): pp. 991–998.

16 Bechara, A., H. Damasio, and A.R. Damasio, 'Emotion, decision making and the orbitofrontal cortex', *Cerebral Cortex*, 2000, 10(3): pp. 295–307.

17 Habib, M., et al., 'Fear and anger have opposite effects on risk seeking in the gain frame', *Frontiers in Psychology*, 2015, 6: p. 253.

18 Harmon-Jones, E., 'Anger and the behavioral approach system', *Personality and Individual Differences*, 2003, 35(5): pp. 995–1005.

19 Habib et al., 'Fear and anger have opposite effects'.

20 Deci, E.L. and A.C. Moller, 'The concept of competence: a starting place for understanding intrinsic motivation and self-determined extrinsic motivation', in *Handbook of Competence and Motivation*, A.J. Elliot and C.S. Dweck (eds) (Guilford Publications, 2005), pp. 579–597.

21 Lepper, M.R., D. Greene, and R.E. Nisbett, 'Undermining children's intrinsic interest with extrinsic reward: a test of the "overjustification" hypothesis', *Journal of Personality and Social Psychology*, 1973, 28(1): pp. 129–137.

22 Clanton Harpine, E., 'Is intrinsic motivation better than extrinsic motivation?', in *Group-Centered Prevention in Mental Health: Theory, Training, and Practice*, E. Clanton Harpine (ed.) (Springer International Publishing, 2015), pp. 87–107.

23 Meyer, D.K. and J.C. Turner, 'Discovering emotion in classroom motivation research', *Educational Psychologist*, 2002, 37(2): pp. 107–114.

24 Blackmore, C., D. Tantam, and E. van Deurzen, 'Evaluation of e-learning outcomes: experience from an online psychotherapy education programme', *Open Learning: The Journal of Open, Distance and e-Learning*, 2008, 23(3): pp. 185–201.

25 Megna, P., 'Better living through dread: medieval ascetics, modern philosophers, and the long history of existential anxiety', *PMLA: Publications of the Modern Language Association of America*, 2015, 130(5): pp. 1285–1301.

26 De Berker, A.O., et al., 'Computations of uncertainty mediate acute stress responses in humans', *Nature Communications*, 2016, 7: p. 10996.

27 Fitzpatrick, M., 'The recollection of anxiety: Kierkegaard as our Socratic occasion to transcend unfreedom', *The Heythrop Journal*, 2014, 55(5): pp. 871–882.

28 Legault, L. and M. Inzlicht, 'Self-determination, self-regulation, and the brain: autonomy improves performance by enhancing neuroaffective responsiveness to self-regulation failure', *Journal of Personality and Social Psychology*, 2013, 105(1): pp. 123–138.

29 Brindley, G., 'The colour of light of very long wavelength', *The Journal of Physiology*, 1955, 130(1): p. 35.

30 Mikellides, B., 'Colour psychology: the emotional effects of colour perception', in *Colour Design*, J. Best (ed.) (Woodhead Publishing, 2012), pp. 105–128.

31 Thoen, H.H., et al., 'A different form of color vision in mantis shrimp', *Science*, 2014, 343(6169): pp. 411–413.

32 Dominy, N.J. and P.W. Lucas, 'Ecological importance of trichromatic vision to primates', *Nature*, 2001, 410(6826): pp. 363–366.

33 Politzer, T., 'Vision is our dominant sense', *Brainline*, URL: https://www.brainline.org/article/vision-our-dominant-sense (accessed 15 April 2018), 2008.

34 Goodale, M.A. and A.D. Milner, 'Separate visual pathways for perception and action', *Trends In Neuroscience*, 1992, 15(1): pp. 20–25.

35 Hupka, R.B., et al., 'The colors of anger, envy, fear, and jealousy: a cross-cultural study', *Journal of Cross-Cultural Psychology*, 1997, 28(2): pp. 156–171.

36 Jin, H.-R., et al., 'Study on physiological responses to color stimulation', *International Association of Societies of Design Research*, 2009: pp. 1969–1979.

37 Fetterman, A.K., M.D. Robinson, and B.P. Meier, 'Anger as "seeing red": evidence for a perceptual association', *Cognition & Emotion*, 2012, 26(8): pp. 1445–1458.

38 Pravossoudovitch, K., et al., 'Is red the colour of danger? Testing an implicit red–danger association', *Ergonomics*, 2014, 57(4): pp. 503–510.

39 Ou, L.-C., et al., 'A study of colour emotion and colour preference. Part I: Colour emotions for single colours', *Color Research & Application*, 2004, 29(3): pp. 232–240.

40 Changizi, M.A., Q. Zhang, and S. Shimojo, 'Bare skin, blood and the evolution of primate colour vision', *Biology Letters*, 2006, 2(2): pp. 217–221.

41 Kienle, A., et al., 'Why do veins appear blue? A new look at an old question', *Applied Optics*, 1996, 35(7): pp. 1151–1160.

42 Re, D.E., et al., 'Oxygenated-blood colour change thresholds for perceived facial redness, health, and attractiveness', *PLOS One*, 2011, 6(3): p. e17859.

43 Changizi et al., 'Bare skin, blood'.

44 Changizi et al., 'Bare skin, blood'.

45 Benitez-Quiroz, C.F., R. Srinivasan, and A.M. Martinez, 'Facial color is an efficient mechanism to visually transmit emotion', *Proceedings of the National Academy of Sciences*, 2018, 115(14): pp. 3581–3586.

46 Stephen, I.D., et al., 'Skin blood perfusion and oxygenation colour affect perceived human health', *PLOS One*, 2009, 4(4): p. e5083.

47 Landgrebe, M., et al., 'Effects of colour exposure on auditory and somatosensory perception – hints for cross-modal plasticity', *Neuroendocrinology Letters*, 2008, 29(4): p. 518.

48 Tan, S.-H. and J. Li, 'Restoration and stress relief benefits of urban park and green space', *Chinese Landscape Architecture*, 2009, 6: pp. 79–82.

49 Lee, K.E., et al., '40-second green roof views sustain attention: the role of micro-breaks in attention restoration', *Journal of Environmental Psychology*, 2015, 42: pp. 182–189.

50 Hill, R.A. and R.A. Barton, 'Psychology: red enhances human performance in contests', *Nature*, 2005. 435(7040): p. 293.

51 Gold, A.L., R.A. Morey, and G. McCarthy, 'Amygdala–prefrontal cortex functional connectivity during threat-induced anxiety and goal distraction', *Biological Psychiatry*, 2015, 77(4): pp. 394–403.

52 Greenlees, I.A., M. Eynon, and R.C. Thelwell, 'Color of soccer goalkeepers' uniforms influences the outcome of penalty kicks', *Perceptual and Motor Skills*, 2013, 117(1): pp. 1–10.

53 Elliot, A.J. and M.A. Maier, 'Color psychology: effects of perceiving color on psychological functioning in humans', *Annual Review of Psychology*, 2014, 65: pp. 95–120.

54 Colombetti, G., 'Appraising valence', *Journal of Consciousness Studies*, 2005, 12(8–9): pp. 103–126.

55 Spence, C., 'Why is piquant/spicy food so popular?' *International Journal of Gastronomy and Food Science*, 2018, 12: pp. 16–21.

56 Frias, B. and A. Merighi, 'Capsaicin, nociception and pain', *Molecules*, 2016, 21(6): p. 797.

57 Omolo, M.A., et al., 'Antimicrobial properties of chili peppers', *Journal of Infectious Diseases and Therapy*, 2014.

58 Rozin, P. and D. Schiller, 'The nature and acquisition of a preference for chili pepper by humans', *Motivation and Emotion*, 1980, 4(1): pp. 77–101.

59 Spence, 'Why is piquant/spicy food so popular?'

60 Hawkes, C., 'Endorphins: the basis of pleasure?' *Journal of Neurology, Neurosurgery & Psychiatry*, 1992, 55(4): pp. 247–250.

61 Solinas, M., S.R. Goldberg, and D. Piomelli, 'The endocannabinoid system in brain reward processes', *British Journal of Pharmacology*, 2008, 154(2): pp. 369–383.

62 Levin, R. and A. Riley, 'The physiology of human sexual function', *Psychiatry*, 2007, 6(3): pp. 90–94.

63 Kawamichi, H., et al., 'Increased frequency of social interaction is associated with enjoyment enhancement and reward system activation', *Scientific Reports*, 2016, 6(1): pp. 1–11.

64 National Institute of Mental Health, 'Human brain appears "hard-wired" for hierarchy', *ScienceDaily*, 2008.

65 Beery, A.K. and D. Kaufer, 'Stress, social behavior, and resilience: insights from rodents', *Neurobiology of Stress*, 2015, 1: pp. 116–127.

66 Wuyts, E., et al., 'Between pleasure and pain: a pilot study on the biological mechanisms associated with BDSM interactions in dominants and submissives', *The Journal of Sexual Medicine*, 2020, 17(4): pp. 784–792.

67 Simula, B.L., 'A "different economy of bodies and pleasures"?: differentiating and evaluating sex and sexual BDSM experiences', *Journal of Homosexuality*, 2019, 66(2): pp. 209–237.

68 Dunkley, C.R., et al., 'Physical pain as pleasure: a theoretical perspective', *The Journal of Sex Research*, 2020, 57(4): pp. 421–437.

69 Vandermeersch, P., 'Self-flagellation in the Early Modern Era', in *The Sense of Suffering: Constructions of Physical Pain in Early Modern Culture*, J.F. van Dijkhuizen and K.A.E. Enenkel (eds) (Brill, 2009), pp. 253–265.

70 Bryant, J. and D. Miron, 'Excitation-transfer theory and three-factor theory of emotion', in *Communication and Emotion*, J. Bryant, D.R. Roskos-Ewoldsen and J. Cantor (eds) (Routledge, 2003), pp. 39–68.

71 McCarthy, D.E., et al., 'Negative reinforcement: possible clinical implications of an integrative model', in, *Substance Abuse and Emotion*, J.D. Kassel (ed.) (American Psychological Association, 2010), pp. 15–42.

72 Raderschall, C.A., R.D. Magrath, and J.M. Hemmi, 'Habituation under natural conditions: model predators are distinguished by approach direction', *Journal of Experimental Biology*, 2011, 214(24): pp. 4209–4216.

73 Krebs, R., et al., 'Novelty increases the mesolimbic functional connectivity of the substantia nigra/ventral tegmental area (SN/VTA) during reward anticipation: evidence from high-resolution fMRI', *NeuroImage*, 2011, 58(2): pp. 647–655.

74 Johnson-Laird, P.N., 'Mental models, deductive reasoning, and the brain', *The Cognitive Neurosciences*, 1995, 65: pp. 999–1008.

75 Finucane, A.M., 'The effect of fear and anger on selective attention', *Emotion*, 2011, 11(4): p. 970.

76 Fredrickson, B.L. and C. Branigan, 'Positive emotions broaden the scope of attention and thought-action repertoires', *Cognition & Emotion*, 2005, 19(3): pp. 313–332.

77 Gasper, K. and G.L. Clore, 'Attending to the big picture: mood and global versus local processing of visual information', *Psychological Science*, 2002, 13(1): pp. 34–40.

78 Melamed, S., et al., 'Attention capacity limitation, psychiatric parameters and their impact on work involvement following brain injury', *Scandinavian Journal of Rehabilitation Medicine*, Supplement, 1985, 12: pp. 21–26.

79 Unkelbach, C., J.P. Forgas, and T.F. Denson, 'The turban effect: the influence of Muslim headgear and induced affect on aggressive responses in the shooter bias paradigm', *Journal of Experimental Social Psychology*, 2008, 44(5): pp. 1409–1413.

80 Spicer, A. and C. Cederström, 'The research we've ignored about happiness at work', *Harvard Business Review*, 21 July 2015.

81 Bless, H. and K. Fiedler, 'Mood and the regulation of information processing and behavior', in *Affect in Social Thinking and Behavior*, J. Forgas (ed.) (Psychology Press, 2006), pp. 65–84.

82 Bless and Fiedler, 'Mood and the regulation of information processing'.

83 Forgas, J.P., 'Don't worry, be sad! On the cognitive, motivational, and interpersonal benefits of negative mood', *Current Directions in Psychological Science*, 2013, 22(3): pp. 225–232.

84 Forgas, J.P., 'Cognitive theories of affect', in *The Corsini Encyclopedia of Psychology*, I.B. Weiner and W.E. Craighead (eds) (John Wiley, 2010), pp. 1–3.

85 Tamir, M., M.D. Robinson, and E.C. Solberg, 'You may worry, but can you recognize threats when you see them? Neuroticism, threat identifications, and negative affect', *Journal of Personality*, 2006, 74(5): pp. 1481–1506.

86 Garcia, E.E., 'Rachmaninoff and Scriabin: creativity and suffering in talent and genius', *The Psychoanalytic Review*, 2004, 91(3): pp. 423–442.

87 Rodriguez, T., 'Negative emotions are key to well-being', *Scientific American*, 2013, 24(2): pp. 26–27.

88 Brown, J.T. and G.A. Stoudemire, 'Normal and pathological grief', *JAMA: the Journal of the American Medical Association*, 1983, 250(3): pp. 378–382.

89 Rachman, S., 'Emotional processing', *Behaviour Research and Therapy*, 1980, 18(1): pp. 51–60.

90 Litz, B.T., et al., 'Emotional processing in posttraumatic stress disorder', *Journal of Abnormal Psychology*, 2000, 109(1): p. 26.

91 Stapleton, J.A., S. Taylor, and G.J. Asmundson, 'Effects of three PTSD treatments on anger and guilt: exposure therapy, eye movement desensitization and reprocessing, and relaxation training', *Journal of Traumatic Stress*, 2006, 19(1): pp. 19–28.

92 Saarni, C., *The Development of Emotional Competence* (Guilford Press, 1999).

93   Shallcross, A.J., et al., 'Let it be: accepting negative emotional experiences predicts decreased negative affect and depressive symptoms', *Behaviour Research and Therapy*, 2010, 48(9): pp. 921–929.

94   Shallcross et al., 'Let it be'.

95   Sharman, L. and G.A. Dingle, 'Extreme metal music and anger processing', *Frontiers in Human Neuroscience*, 2015, 9: p. 272.

96   Tamir, M. and Y. Bigman, 'Why might people want to feel bad? Motives in contrahedonic emotion regulation', in *The Positive Side of Negative Emotions*, W. Gerrod Parrott (ed.) (Guilford Press, 2014), pp. 201–223.

97   Saraiva, A.C., F. Schüür, and S. Bestmann, 'Emotional valence and contextual affordances flexibly shape approach-avoidance movements', *Frontiers in Psychology*, 2013, 4: p. 933.

98   Snyder, M. and A. Frankel, 'Observer bias: a stringent test of behavior engulfing the field', *Journal of Personality and Social Psychology*, 1976, 34: pp. 857–864.

99   Karanicolas, P.J., F. Farrokhyar, and M. Bhandari, 'Blinding: who, what, when, why, how?' *Canadian Journal of Surgery*, 2010, 53(5): p. 345.

100   Burghardt, G.M., et al., 'Perspectives – minimizing observer bias in behavioral studies: a review and recommendations', *Ethology*, 2012, 118(6): pp. 511–517.

101   Dvorsky, G., 'The neuroscience of stage fright – and how to cope with it', *Gizmodo*, 10 October 2012.

102   Wesner, R.B., R. Noyes Jr, and T.L. Davis, 'The occurrence of performance anxiety among musicians', *Journal of Affective Disorders*, 1990, 18(3): pp. 177–185.

103   Chao-Gang, W., 'Through theory of the two brain hemispheres' work division to look for the solution of stage fright problem – an inspiration of tennis ball movement in heart', *Journal of Xinghai Conservatory of Music*, 2003(2): p. 6.

104   Toda, T., et al., 'The role of adult hippocampal neurogenesis in brain health and disease', *Molecular Psychiatry*, 2019, 24(1): pp. 67–87.

105   Teigen, K.H., 'Yerkes-Dodson: a law for all seasons', *Theory & Psychology*, 1994, 4(4): pp. 525–547.

106   Kawamichi, et al. 'Increased frequency of social interaction'.

107   Kross, E., et al., 'Social rejection shares somatosensory representations with physical pain', *Proceedings of the National Academy of Sciences*, 2011, 108(15): pp. 6270–6275.

108   Trower, P. and P. Gilbert, 'New theoretical conceptions of social anxiety and social phobia', *Clinical Psychology Review*, 1989, 9(1): pp. 19–35.

109 Dvorsky, 'The neuroscience of stage fright'.
110 Kotov, R., et al., 'Personality traits and anxiety symptoms: the multilevel trait predictor model', *Behaviour Research and Therapy*, 2007, 45(7): pp. 1485–1503.
111 Nagel, J.J., 'Stage fright in musicians: a psychodynamic perspective', *Bulletin of the Menninger Clinic*, 1993, 57(4): p. 492.
112 McRae, R.R., et al., 'Sources of structure: genetic, environmental, and artifactual influences on the covariation of personality traits', *Journal of Personality*, 2001, 69(4): pp. 511–535.
113 Nagel, 'Stage fright in musicians'.
114 Holmes, J., 'Attachment theory', in *The Wiley-Blackwell Encyclopedia of Social Theory*, B.S Turner et al. (eds) (Wiley-Blackwell, 2017), pp. 1–3.
115 Brooks, A.W., 'Get excited: reappraising pre-performance anxiety as excitement', *Journal of Experimental Psychology: General*, 2014, 143(3): p. 1144.
116 Denton, D.A., et al., 'The role of primordial emotions in the evolutionary origin of consciousness', *Consciousness and Cognition*, 2009, 18(2): pp. 500–514.
117 Ferrier, D.E., H.H. Bassett, and S.A. Denham, 'Relations between executive function and emotionality in preschoolers: exploring a transitive cognition–emotion linkage', *Frontiers in Psychology*, 2014, 5: p. 487.
118 Rueda, M.R. and P. Paz-Alonzo, 'Executive function and emotional development', *Contexts*, 2013, 1: p. 2.
119 Campos, J.J., C.B. Frankel, and L. Camras, 'On the nature of emotion regulation', *Child Development*, 2004, 75(2): pp. 377–394.
120 Davidson, 'Well-being and affective style'.
121 Jumah, F.R. and R.H. Dossani, 'Neuroanatomy, Cingulate Cortex', in *StatPearls [Internet]* (StatPearls Publishing, 2019).
122 Shackman, A.J., et al., 'The integration of negative affect, pain and cognitive control in the cingulate cortex', *Nature Reviews Neuroscience*, 2011, 12(3): pp. 154–167.
123 Etkin, A., T. Egner, and R. Kalisch, 'Emotional processing in anterior cingulate and medial prefrontal cortex', *Trends in Cognitive Sciences*, 2011, 15(2): pp. 85–93.
124 Sobol, I. and Y.L. Levitan, 'A pseudo-random number generator for personal computers', *Computers & Mathematics with Applications*, 1999, 37(4–5): pp. 33–40.

3: Emotional Memories

1   Burnett, D.J., 'Role of the hippocampus in configural learning', PhD thesis, 2010, Cardiff University.

2   Christianson, S.-Å., 'Remembering emotional events: potential mechanisms', in *The Handbook of Emotion and Memory: Research and Theory*, S.-Å. Christianson (ed.) (Psychology Press, 1992), pp. 307–340.

3   Gailene, D., V. Lepeshkene, and A. Shiurkute, 'Features of the "Zeigarnik effect" in psychiatric clinical practice', *Zhurnal nevropatologii i psikhiatrii imeni SS Korsakova*, 1980, 80(12): pp. 1837–1841.

4   Tulving, E., 'How many memory systems are there?' *American Psychologist*, 1985, 40(4): p. 385.

5   Nagao, S. and H. Kitazawa, 'Role of the cerebellum in the acquisition and consolidation of motor memory', *Brain and nerve = Shinkei kenkyu no shinpo*, 2008, 60(7): pp. 783–790.

6   Pessiglione, M., et al., 'Subliminal instrumental conditioning demonstrated in the human brain', *Neuron*, 2008, 59(4): pp. 561–567.

7   Turner, B.M., et al., 'The cerebellum and emotional experience', *Neuropsychologia*, 2007, 45(6): pp. 1331–1341.

8   Cardinal, R.N., et al., 'Emotion and motivation: the role of the amygdala, ventral striatum, and prefrontal cortex', *Neuroscience & Biobehavioral Reviews*, 2002, 26(3): pp. 321–352.

9   Squire, L.R. and B.J. Knowlton, 'Memory, hippocampus, and brain systems', in *The Cognitive Neurosciences*, M.S. Gazzaniga (ed.) (MIT Press, 1995), pp. 825–837.

10 Buckner, R.L. and S.E. Petersen, 'What does neuroimaging tell us about the role of prefrontal cortex in memory retrieval?' *Seminars in Neuroscience*, 1996, 8(1): pp. 47–55.

11 Squire, L.R. and B.J. Knowlton, 'The medial temporal lobe, the hippocampus, and the memory systems of the brain', *The New Cognitive Neurosciences*, 2000, 2: pp. 756–776.

12 Mayford, M., S.A. Siegelbaum, and E.R. Kandel, 'Synapses and memory storage', *Cold Spring Harbor Perspectives in Biology*, 2012, 4(6): p. a005751.

13 Toda, et al., 'The role of adult hippocampal neurogenesis'.

14 Phelps, E.A., 'Human emotion and memory: interactions of the amygdala and hippocampal complex', *Current Opinion in Neurobiology*, 2004, 14(2): pp. 198–202.

15 Amaral, D.G., H. Behniea, and J.L. Kelly, 'Topographic organization of projections from the amygdala to the visual cortex in the macaque monkey', *Neuroscience*, 2003, 118(4): pp. 1099–1120.

16 Öhman, A., A. Flykt, and F. Esteves, 'Emotion drives attention: detecting the snake in the grass', *Journal of Experimental Psychology: General*, 2001, 130(3): p. 466.

17 Ben-Haim, M.S., et al., 'The emotional Stroop task: assessing cognitive performance under exposure to emotional content', *JoVE (Journal of Visualized Experiments)*, 2016, 112: p. e53720.

18 Talarico, J.M., D. Berntsen, and D.C. Rubin, 'Positive emotions enhance recall of peripheral details', *Cognition & Emotion*, 2009, 23(2): pp. 380–398.

19 Phelps, 'Human emotion and memory'.

20 White, A.M., 'What happened? Alcohol, memory blackouts, and the brain', *Alcohol Research & Health*, 2003, 27(2): p. 186.

21 Dolcos, F., K.S. LaBar, and R. Cabeza, 'Interaction between the amygdala and the medial temporal lobe memory system predicts better memory for emotional events', *Neuron*, 2004, 42(5): pp. 855–863.

22 Oakes, M. and R. Bor, 'The psychology of fear of flying (part I): a critical evaluation of current perspectives on the nature, prevalence and etiology of fear of flying', *Travel Medicine and Infectious Disease*, 2010, 8(6): pp. 327–338.

23 Phelps, E.A., et al., 'Activation of the left amygdala to a cognitive representation of fear', *Nature Neuroscience*, 2001, 4(4): pp. 437–441.

24 McGaugh, J.L., 'Memory – a century of consolidation', *Science*, 2000, 287(5451): pp. 248–251.

25 Phelps, 'Human emotion and memory'.

26 McKay, L. and J. Cidlowski, 'Pharmacokinetics of corticosteroids', in *Holland-Frei Cancer Medicine*, Sixth edn, D.W. Kufe et al. (eds) (BC Decker, 2003).

27 McGaugh, 'Memory'.

28 Dunsmoor, J.E., et al., 'Emotional learning selectively and retroactively strengthens memories for related events', *Nature*, 2015, 520(7547): pp. 345–348.

29 Mercer, T., 'Wakeful rest alleviates interference-based forgetting', *Memory*, 2015, 23(2): pp. 127–137.

30 Akers, K.G., et al., 'Hippocampal neurogenesis regulates forgetting during adulthood and infancy', *Science*, 2014, 344(6184): pp. 598–602.

31 Davis, R.L. and Y. Zhong, 'The biology of forgetting—a perspective', *Neuron*, 2017, 95(3): pp. 490–503.

32   Sherman, E., 'Reminiscentia: cherished objects as memorabilia in late-life reminiscence', *The International Journal of Aging and Human Development*, 1991, 33(2): pp. 89–100.

33   Sherman, 'Reminiscentia'.

34   Levy, B.J. and M.C. Anderson, 'Inhibitory processes and the control of memory retrieval', *Trends in Cognitive Sciences*, 2002, 6(7): pp. 299–305.

35   Brown and Stoudemire, 'Normal and pathological grief'.

36   Bridge, D.J. and J.L. Voss, 'Hippocampal binding of novel information with dominant memory traces can support both memory stability and change', *Journal of Neuroscience*, 2014, 34(6): pp. 2203–2213.

37   Skowronski, J.J., 'The positivity bias and the fading affect bias in autobiographical memory', in *Handbook of Self-enhancement and Self-protection*, M.D. Alicke and C. Sedikides (eds) (Guilford Press, 2011), p. 211.

38   Rozin, P. and E.B. Royzman, 'Negativity bias, negativity dominance, and contagion', *Personality and Social Psychology Review*, 2001, 5(4): pp. 296–320.

39   Vaish, A., T. Grossmann, and A. Woodward, 'Not all emotions are created equal: the negativity bias in social-emotional development', *Psychological Bulletin*, 2008, 134(3): pp. 383–403.

40   Gibbons, J.A., S.A. Lee, and W.R. Walker, 'The fading affect bias begins within 12 hours and persists for 3 months', *Applied Cognitive Psychology*, 2011, 25(4): pp. 663–672.

41   Walker, W.R., et al., 'On the emotions that accompany autobiographical memories: dysphoria disrupts the fading affect bias', *Cognition and Emotion*, 2003, 17(5): pp. 703–723.

42   Croucher, C.J., et al., 'Disgust enhances the recollection of negative emotional images', *PLOS One*, 2011, 6(11): p. e26571.

43   Tybur, J.M., et al., 'Disgust: evolved function and structure', *Psychological Review*, 2013, 120(1): p. 65.

44   Konnikova, M., 'Smells like old times', *Scientific American Mind*, 2012, 23(1): pp. 58–63.

45   Politzer, 'Vision is our dominant sense'.

46   Zeng, F.-G., Q.-J. Fu, and R. Morse, 'Human hearing enhanced by noise', *Brain research*, 2000, 869(1–2): pp. 251–255.

47   Vassar, R., J. Ngai, and R. Axel, 'Spatial segregation of odorant receptor expression in the mammalian olfactory epithelium', *Cell*, 1993, 74(2): pp. 309–318.

48 Shepherd, G.M. and C.A. Greer, 'Olfactory bulb', in *The Synaptic Organization of the Brain*, G.M. Shepherd (ed.) (Oxford University Press, 1998), pp. 159–203.

49 Soudry, Y., et al., 'Olfactory system and emotion: common substrates', *European Annals of Otorhinolaryngology, Head and Neck Diseases*, 2011, 128(1): pp. 18–23.

50 Rowe, T.B., T.E. Macrini, and Z.-X. Luo, 'Fossil evidence on origin of the mammalian brain', *Science*, 2011, 332(6032): pp. 955–957.

51 Eichenbaum, H., 'The role of the hippocampus in navigation is memory', *Journal of Neurophysiology*, 2017, 117(4): pp. 1785–1796.

52 Maguire, E.A., R.S. Frackowiak, and C.D. Frith, 'Recalling routes around London: activation of the right hippocampus in taxi drivers', *Journal of Neuroscience*, 1997, 17(18): pp. 7103–7110.

53 Kumaran, D. and E.A. Maguire, 'The human hippocampus: cognitive maps or relational memory?' *Journal of Neuroscience*, 2005, 25(31): pp. 7254–7259.

54 Aboitiz, F. and J.F. Montiel, 'Olfaction, navigation, and the origin of isocortex', *Frontiers in Neuroscience*, 2015, 9(402).

55 Pedersen, P.E., et al., 'Evidence for olfactory function in utero', *Science*, 1983, 221(4609): pp. 478–480.

56 Vantoller, S. and M. Kendalreed, 'A possible protocognitive role for odor in human infant development', *Brain and Cognition*, 1995, 29(3): pp. 275–293.

57 Willander, J. and M. Larsson, 'Smell your way back to childhood: autobiographical odor memory', *Psychonomic Bulletin & Review*, 2006, 13(2): pp. 240–244.

58 Yeshurun, Y., et al., 'The privileged brain representation of first olfactory associations', *Current Biology*, 2009, 19(21): pp. 1869–1874.

59 Hwang, K., et al., 'The human thalamus is an integrative hub for functional brain networks', *Journal of Neuroscience*, 2017, 37(23): pp. 5594–5607.

60 Rowe et al., 'Fossil evidence'.

61 Aqrabawi, A.J. and J.C. Kim, 'Hippocampal projections to the anterior olfactory nucleus differentially convey spatiotemporal information during episodic odour memory', *Nature Communications*, 2018, 9(1): pp. 1–10.

62 Soudry, et al., 'Olfactory system and emotion'.

63 De Araujo, I.E., et al., 'Taste-olfactory convergence, and the representation of the pleasantness of flavour, in the human brain', *European Journal of Neuroscience*, 2003, 18(7): pp. 2059–2068.

64 Weber, S.T. and E. Heuberger, 'The impact of natural odors on affective states in humans', *Chemical Senses*, 2008, 33(5): pp. 441–447.

65 Herz, R.S. and J. von Clef, 'The influence of verbal labeling on the perception of odors: evidence for olfactory illusions?' *Perception*, 2001, 30(3): pp. 381–391.

66 Chen, D. and J. Haviland-Jones, 'Human olfactory communication of emotion', *Perceptual and Motor Skills*, 2000, 91(3): pp. 771–781.

67 Zald, D.H. and J.V. Pardo, 'Emotion, olfaction, and the human amygdala: amygdala activation during aversive olfactory stimulation', *Proceedings of the National Academy of Sciences*, 1997, 94(8): pp. 4119–4124.

68 Soudry, et al., 'Olfactory system and emotion'.

69 Deliberto, T., 'The first and ultimate primary emotion – fear', in *The Psychology Easel*, 2011, Blogspot.com: http://taradeliberto.blogspot.com/2011/03/first-emotion-fear.html.

70 Willander, J. and M. Larsson, 'Olfaction and emotion: the case of autobiographical memory', *Memory & Cognition*, 2007, 35(7): pp. 1659–1663.

71 Konnikova, 'Smells like old times'.

72 Taalman, H., C. Wallace, and R. Milev, 'Olfactory functioning and depression: a systematic review', *Frontiers in Psychiatry*, 2017, 8: p. 190.

73 Tukey, A., 'Notes on involuntary memory in Proust', *The French Review*, 1969, 42(3): pp. 395–402.

74 Juslin, P.N. and D. Västfjäll, 'Emotional responses to music: the need to consider underlying mechanisms', *Behavioral and Brain Sciences*, 2008, 31(5): pp. 559–575.

75 Skoe, E. and N. Kraus, 'Auditory brainstem response to complex sounds: a tutorial', *Ear and Hearing*, 2010, 31(3): p. 302.

76 Raizada, R.D. and R.A. Poldrack, 'Challenge-driven attention: interacting frontal and brainstem systems', *Frontiers in Human Neuroscience*, 2008, 2: p. 3.

77 Burt, J.L., et al., 'A psychophysiological evaluation of the perceived urgency of auditory warning signals', *Ergonomics*, 1995, 38(11): pp. 2327–2340.

78 Nozaradan, S., I. Peretz, and A. Mouraux, 'Selective neuronal entrainment to the beat and meter embedded in a musical rhythm', *Journal of Neuroscience*, 2012, 32(49): pp. 17572–17581.

79 DeNora, T., 'Aesthetic agency and musical practice: new directions in the sociology of music and emotion', in *Music and Emotion: Theory and Research*, P.N. Juslin and J.A. Sloboda (eds) (Oxford University Press, 2001), pp. 161–180.

80 Juslin and Västfjäll, 'Emotional responses to music'.

81 Deliège, I. and J.A. Sloboda, *Musical Beginnings: Origins and Development of Musical Competence* (Oxford University Press, 1996).

82 Egermann, H. and S. McAdams, 'Empathy and emotional contagion as a link between recognized and felt emotions in music listening', *Music Perception: An Interdisciplinary Journal*, 2012, 31(2): pp. 139–156.

83 Di Pellegrino, G., et al., 'Understanding motor events: a neurophysiological study', *Experimental Brain Research*, 1992, 91(1): pp. 176–180.

84 Kilner, J.M. and R.N. Lemon, 'What we know currently about mirror neurons', *Current Biology*, 2013, 23(23): pp. R1057–R1062.

85 Acharya, S. and S. Shukla, 'Mirror neurons: enigma of the metaphysical modular brain', *Journal of Natural Science, Biology, and Medicine*, 2012, 3(2): p. 118.

86 Engelen, T., et al., 'A causal role for inferior parietal lobule in emotion body perception', *Cortex*, 2015, 73: pp. 195–202.

87 Decety, J. and P.L. Jackson, 'The functional architecture of human empathy', *Behavioral and Cognitive Neuroscience Reviews*, 2004, 3(2): pp. 71–100.

88 Gazzola, V., L. Aziz-Zadeh, and C. Keysers, 'Empathy and the somatotopic auditory mirror system in humans', *Current Biology*, 2006, 16(18): pp. 1824–1829.

89 Huron, D. and E.H. Margulis, 'Musical expectancy and thrills', in *Handbook of Music and Emotion: Theory, Research, Applications*, P.N. Juslin and J.A. Sloboda (eds), (Oxford University Press, 2010), pp. 575–604.

90 Patel, A.D., 'Language, music, syntax and the brain', *Nature Neuroscience*, 2003, 6(7): pp. 674–681.

91 Krumhansl, C.L., et al., 'Melodic expectation in Finnish spiritual folk hymns: convergence of statistical, behavioral, and computational approaches', *Music Perception: An Interdisciplinary Journal*, 1999, 17(2): pp. 151–195.

92 Patel, 'Language, music, syntax'.

93 Partanen, E., et al., 'Prenatal music exposure induces long-term neural effects', *PLOS One*, 2013, 8(10).

94 Pereira, C.S., et al., 'Music and emotions in the brain: familiarity matters', *PLOS One*, 2011, 6(11).

95 Burwell, R.D., 'The parahippocampal region: corticocortical connectivity', *Annals – New York Academy of Sciences*, 2000, 911: pp. 25–42.

96 Caruana, F., et al., 'Motor and emotional behaviours elicited by electrical stimulation of the human cingulate cortex, *Brain*, 2018, 141(10): pp. 3035–3051.

97 Hofmann, W., et al., 'Evaluative conditioning in humans: a meta-analysis', *Psychological Bulletin*, 2010, 136(3): p. 390.

98   Balleine, B.W. and S. Killcross, 'Parallel incentive processing: an integrated view of amygdala function', *Trends in Neurosciences*, 2006, 29(5): pp. 272–279.

99   Sacchetti, B., B. Scelfo, and P. Strata, 'The cerebellum: synaptic changes and fear conditioning', *The Neuroscientist*, 2005, 11(3): pp. 217–227.

100   Juslin and Västfjäll, 'Emotional responses to music'.

101   LeDoux, J.E., 'Emotion: clues from the brain', *Annual Review of Psychology*, 1995, 46(1): pp. 209–235.

102   Gabrielsson, A., 'Emotion perceived and emotion felt: same or different?' *Musicae Scientiae*, 2001, 5(1_suppl): pp. 123–147.

103   Lang, P.J., 'A bio-informational theory of emotional imagery', *Psychophysiology*, 1979, 16(6): pp. 495–512.

104   Tingley, J., M. Moscicki and K. Buro, 'The effect of earworms on affect', *MacEwan University Student Research Proceedings*, 2019, 4(2).

105   Singhal, D., 'Why this Kolaveri Di: maddening phenomenon of earworm', 2011. Available at SSRN 1969781.

106   Schulkind, M.D., L.K. Hennis, and D.C. Rubin, 'Music, emotion, and autobiographical memory: they're playing your song', *Memory & Cognition*, 1999, 27(6): pp. 948–955.

107   Rathbone, C.J., C.J. Moulin, and M.A. Conway, 'Self-centered memories: the reminiscence bump and the self,' *Memory & Cognition*, 2008, 36(8): pp. 1403–1414.

108   Mills, K.L., et al., 'The developmental mismatch in structural brain maturation during adolescence', *Developmental Neuroscience*, 2014, 36(3–4): pp. 147–160.

109   Blood, A.J. and R.J. Zatorre, 'Intensely pleasurable responses to music correlate with activity in brain regions implicated in reward and emotion', *Proceedings of the National Academy of Sciences*, 2001, 98(20): pp. 11818–11823.

110   Boero, D.L. and L. Bottoni, 'Why we experience musical emotions: intrinsic musicality in an evolutionary perspective', *Behavioral and Brain Sciences*, 2008, 31(5): pp. 585–586.

111   Simpson, E.A., W.T. Oliver, and D. Fragaszy, 'Super-expressive voices: music to my ears?' *Behavioral and Brain Sciences*, 2008, 31(5): pp. 596–597.

112   Simpson et al., 'Super-expressive voices'.

113   Krach, S., et al., 'The rewarding nature of social interactions', *Frontiers in Behavioral Neuroscience*, 2010, 4: p. 22.

114 Alcorta, C.S., R. Sosis, and D. Finkel, 'Ritual harmony: toward an evolutionary theory of music', *Behavioral and Brain Sciences*, 2008, 31(5): pp. 576–577.

115 Freeman, W.J., 'Happiness doesn't come in bottles. Neuroscientists learn that joy comes through dancing, not drugs', *Journal of Consciousness Studies*, 1997, 4(1): pp. 67–70.

116 Krakauer, J., 'Why do we like to dance – and move to the beat', *Scientific American*, 26 September 2008.

117 Peery, J.C., I.W. Peery, and T.W. Draper, *Music and Child Development* (Springer Science & Business Media, 2012).

118 Levin, R., 'Sleep and dreaming characteristics of frequent nightmare subjects in a university population', *Dreaming*, 1994, 4(2): pp. 127–137.

119 National Institute of Neurological Disorders and Stroke, *Brain Basics: Understanding Sleep* (NINDS, 2006).

120 Kaufman, D.M., H.L. Geyer, and M.J. Milstein, 'Sleep disorders', in *Kaufman's Clinical Neurology for Psychiatrists*, Eighth edn, D.M. Kaufman, H.L. Geyer, and M.J. Milstein (eds) (Elsevier, 2017), pp. 361–388.

121 Wamsley, E.J., 'Dreaming and offline memory consolidation', *Current Neurology and Neuroscience Reports*, 2014, 14(3): p. 433.

122 Nielsen, T.A. and P. Stenstrom, 'What are the memory sources of dreaming?' *Nature*, 2005, 437(7063): pp. 1286–1289.

123 Smith, K., 'Rose-scented sleep improves memory', *Nature*, 8 March 2007.

124 Walker, M.P., et al., 'Cognitive flexibility across the sleep–wake cycle: REM-sleep enhancement of anagram problem solving', *Cognitive Brain Research*, 2002, 14(3): pp. 317–324.

125 Schredl, M. and F. Hofmann, 'Continuity between waking activities and dream activities', *Consciousness and Cognition*, 2003, 12(2): pp. 298–308.

126 Braun, A.R., et al., 'Dissociated pattern of activity in visual cortices and their projections during human rapid eye movement sleep, *Science*, 1998, 279(5347): pp. 91–95.

127 Nielsen and Stenstrom, 'What are the memory sources of dreaming?'.

128 Freud, S. and J. Strachey, *The Interpretation of Dreams* (Gramercy Books, 1996).

129 Nielsen, T. and R. Levin, 'Nightmares: a new neurocognitive model', *Sleep Medicine Reviews*, 2007, 11(4): pp. 295–310.

130 Nielsen and Stenstrom, 'What are the memory sources of dreaming?'.

131 Popp, C.A., et al., 'Repetitive relationship themes in waking narratives and dreams', *Journal of Consulting and Clinical Psychology*, 1996, 64(5): p. 1073.

132 Revonsuo, A., 'The reinterpretation of dreams: an evolutionary hypothesis of the function of dreaming', *Behavioral and Brain Sciences*, 2000, 23(6): pp. 877–901.

133 Fisher, B.E., C. Pauley, and K. McGuire, 'Children's sleep behavior scale: normative data on 870 children in grades 1 to 6', *Perceptual and Motor Skills*, 1989, 68(1): pp. 227–236.

134 Levin, R. and T.A. Nielsen, 'Disturbed dreaming, posttraumatic stress disorder, and affect distress: a review and neurocognitive model', *Psychological Bulletin*, 2007, 133(3): pp. 482–528.

135 Langston, T.J., J.L. Davis, and R.M. Swopes, 'Idiopathic and posttrauma nightmares in a clinical sample of children and adolescents: characteristics and related pathology', *Journal of Child & Adolescent Trauma*, 2010, 3(4): pp. 344–356.

136 Brown, R.J. and D.C. Donderi, 'Dream content and self-reported well-being among recurrent dreamers, past-recurrent dreamers, and nonrecurrent dreamers', *Journal of Personality and Social Psychology*, 1986, 50(3): p. 612.

137 Quirk, G.J., 'Memory for extinction of conditioned fear is long-lasting and persists following spontaneous recovery', *Learning & Memory*, 2002, 9(6): pp. 402–407.

138 Spoormaker, V.I., M. Schredl, and J. van den Bout, 'Nightmares: from anxiety symptom to sleep disorder', *Sleep Medicine Reviews*, 2006, 10(1): pp. 19–31.

4: Emotional Communication

1 McHenry, M., et al., 'Voice analysis during bad news discussion in oncology: reduced pitch, decreased speaking rate, and nonverbal communication of empathy', *Supportive Care in Cancer*, 2012, 20(5): pp. 1073–1078.

2 Kana, R.K. and B.G. Travers, 'Neural substrates of interpreting actions and emotions from body postures', *Social Cognitive and Affective Neuroscience*, 2012, 7(4): pp. 446–456.

3 Book, A., K. Costello, and J.A. Camilleri, 'Psychopathy and victim selection: the use of gait as a cue to vulnerability', *Journal of Interpersonal Violence*, 2013, 28(11): pp. 2368–2383.

4 Scott, S.K., et al., 'The social life of laughter', *Trends in Cognitive Sciences*, 2014, 18(12): pp. 618–620.

5 Seyfarth, R.M. and D.L. Cheney, 'Affiliation, empathy, and the origins of theory of mind', *Proceedings of the National Academy of Sciences*, 2013, 110(Supplement 2): pp. 10349–10356.

6   Levinson, S.C., 'Spatial cognition, empathy and language evolution', *Studies in Pragmatics*, 2018, 20: pp. 16–21.

7   Land, W., et al., 'From action representation to action execution: exploring the links between cognitive and biomechanical levels of motor control', *Frontiers in Computational Neuroscience*, 2013, 7: p. 127.

8   Meltzoff, A.N. and M.K. Moore, 'Persons and representation: why infant imitation is important for theories of human development', in *Imitation in Infancy*, J. Nadel and G. Butterworth (eds) (Cambridge University Press, 1999), pp. 9–35.

9   Carr, L., et al., 'Neural mechanisms of empathy in humans: a relay from neural systems for imitation to limbic areas', *Proceedings of the National Academy of Sciences*, 2003, 100(9): pp. 5497–5502.

10 Karnath, H.-O., 'New insights into the functions of the superior temporal cortex', *Nature Reviews Neuroscience*, 2001, 2(8): pp. 568–576.

11 Andersen, R.A. and C.A. Buneo, 'Intentional maps in posterior parietal cortex', *Annual Review of Neuroscience*, 2002, 25(1): pp. 189–220.

12 Hartwigsen, G., et al., 'Functional segregation of the right inferior frontal gyrus: evidence from coactivation-based parcellation', *Cerebral Cortex*, 2019, 29(4): pp. 1532–1546.

13 Aron, A.R., T.W. Robbins, and R.A. Poldrack, 'Inhibition and the right inferior frontal cortex: one decade on', *Trends in Cognitive Sciences*, 2014, 18(4): pp. 177–185.

14 Meltzoff and Moore, 'Persons and representation'.

15 Jabbi, M., J. Bastiaansen, and C. Keysers, 'A common anterior insula representation of disgust observation, experience and imagination shows divergent functional connectivity pathways', *PLOS One*, 2008, 3(8): p. e2939.

16 Augustine, J.R., 'Circuitry and functional aspects of the insular lobe in primates including humans', *Brain Research Reviews*, 1996, 22(3): pp. 229–244.

17 Carr, et al., 'Neural mechanisms of empathy in humans'.

18 Eres, R., et al., 'Individual differences in local gray matter density are associated with differences in affective and cognitive empathy', *NeuroImage*, 2015, 117: pp. 305–310.

19 Riess, H., 'The science of empathy', *Journal of Patient Experience*, 2017, 4(2): pp. 74–77.

20 Trevarthen, C., 'Communication and cooperation in early infancy: a description of primary intersubjectivity', *Before Speech: The Beginning of Interpersonal Communication*, 1979, 1: pp. 530–571.

21　Martin, G.B. and R.D. Clark, 'Distress crying in neonates: species and peer specificity', *Developmental Psychology*, 1982, 18(1): p. 3.

22　Van Baaren, R., et al., 'Where is the love? The social aspects of mimicry', *Philosophical Transactions of the Royal Society B: Biological Sciences*, 2009, 364(1528): pp. 2381–2389.

23　Van Baaren, R.B., et al., 'Mimicry and prosocial behavior', *Psychological Science*, 2004, 15(1): pp. 71–74.

24　Chartrand, T.L. and J.A. Bargh, 'The chameleon effect: the perception-behavior link and social interaction', *Journal of Personality and Social Psychology*, 1999, 76(6): pp. 893–910.

25　Maddux, W.W., E. Mullen, and A.D. Galinsky, 'Chameleons bake bigger pies and take bigger pieces: strategic behavioral mimicry facilitates negotiation outcomes', *Journal of Experimental Social Psychology*, 2008, 44(2): pp. 461–468.

26　Book, A., et al., 'The mask of sanity revisited: psychopathic traits and affective mimicry', *Evolutionary Psychological Science*, 2015, 1(2): pp. 91–102.

27　Jackson, P.L., P. Rainville, and J. Decety, 'To what extent do we share the pain of others? Insight from the neural bases of pain empathy', *Pain*, 2006, 125(1): pp. 5–9.

28　Avenanti, A., et al., 'Transcranial magnetic stimulation highlights the sensorimotor side of empathy for pain', *Nature Neuroscience*, 2005, 8(7): pp. 955–960.

29　Nagasako, E.M., A.L. Oaklander, and R.H. Dworkin, 'Congenital insensitivity to pain: an update', *Pain*, 2003, 101(3): pp. 213–219.

30　Danziger, N., K.M. Prkachin, and J.C. Willer, 'Is pain the price of empathy? The perception of others' pain in patients with congenital insensitivity to pain', *Brain*, 2006, 129(9): pp. 2494–2507.

31　Rives Bogart, K. and D. Matsumoto, 'Facial mimicry is not necessary to recognize emotion: facial expression recognition by people with Moebius syndrome', *Social Neuroscience*, 2010, 5(2): pp. 241–251.

32　Watanabe, S. and Y. Kosaki, 'Evolutionary origin of empathy and inequality aversion', in *Evolution of the Brain, Cognition, and Emotion in Vertebrates*, S. Watanabe, M. Hofman, and T. Shimizu (eds) (Springer, 2017), pp. 273–299.

33　De Waal, F.B., 'Putting the altruism back into altruism: the evolution of empathy', *Annual Review of Psychology*, 2008, 59: pp. 279–300.

34　Schroeder, D.A., et al., 'Empathic concern and helping behavior: egoism or altruism?' *Journal of Experimental Social Psychology*, 1988, 24(4): pp. 333–353.

35 Buck, R., 'Communicative genes in the evolution of empathy and altruism', *Behavior Genetics*, 2011, 41(6): pp. 876–888.

36 Stietz, J., et al., 'Dissociating empathy from perspective-taking: evidence from intra- and inter-individual differences research', *Frontiers in Psychiatry*, 2019, 10: p. 126.

37 Batson, C.D., et al., 'Empathic joy and the empathy-altruism hypothesis', *Journal of Personality and Social Psychology*, 1991, 61(3): p. 413.

38 Carr et al., 'Neural mechanisms of empathy in humans'.

39 Gallagher, H.L. and C.D. Frith, 'Functional imaging of "theory of mind"', *Trends in Cognitive Sciences*, 2003, 7(2): pp. 77–83.

40 Allman, J.M., et al., 'The anterior cingulate cortex: the evolution of an interface between emotion and cognition', *Annals of the New York Academy of Sciences*, 2001, 935(1): pp. 107–117.

41 Decety and Jackson, 'The functional architecture of human empathy'.

42 De Vignemont, F. and T. Singer, 'The empathic brain: how, when and why?', *Trends in Cognitive Sciences*, 2006, 10(10): pp. 435–441.

43 Hatfield, E., J.T. Cacioppo, and R.L. Rapson, 'Emotional contagion', *Current Directions in Psychological Science*, 1993, 2(3): pp. 96–100.

44 Hatfield, E., R.L. Rapson, and Y.-C.L. Le, 'Emotional contagion and empathy', in *The Social Neuroscience of Empathy*, J. Decety and W. Ickes (eds) (MIT Press, 2011), p. 19.

45 Schürmann, M., et al., 'Yearning to yawn: the neural basis of contagious yawning', *NeuroImage*, 2005, 24(4): pp. 1260–1264.

46 Guggisberg, A.G., et al., 'Why do we yawn?' *Neuroscience & Biobehavioral Reviews*, 2010, 34(8): pp. 1267–1276.

47 Dunbar, R.I., 'The social brain hypothesis and its implications for social evolution', *Annals of Human Biology*, 2009, 36(5): pp. 562–572.

48 Dolcos, F., A.D. Iordan, and S. Dolcos, 'Neural correlates of emotion–cognition interactions: a review of evidence from brain imaging investigations', *Journal of Cognitive Psychology*, 2011, 23(6): pp. 669–694.

49 Paulson, O.B., et al., 'Cerebral blood flow response to functional activation', *Journal of Cerebral Blood Flow & Metabolism*, 2010, 30(1): pp. 2–14.

50 Ibrahim, J.K., et al., 'State laws restricting driver use of mobile communications devices: distracted-driving provisions, 1992–2010', *American Journal of Preventive Medicine*, 2011, 40(6): pp. 659–665.

51 Stietz, et al., 'Dissociating empathy from perspective-taking'.

52 Dolcos, et al., 'Neural correlates of emotion–cognition interactions'.

53 Vilanova, F., et al., 'Deindividuation: from Le Bon to the social identity model of deindividuation effects', *Cogent Psychology*, 2017, 4(1): p. 1308104.

54 Christoff, K. and J.D.E. Gabrieli, 'The frontopolar cortex and human cognition: evidence for a rostrocaudal hierarchical organization within the human prefrontal cortex,' *Psychobiology*, 2000, 28(2): pp. 168–186.

55 Tong, E.M., D.H. Tan, and Y.L. Tan, 'Can implicit appraisal concepts produce emotion-specific effects? A focus on unfairness and anger', *Consciousness and Cognition*, 2013, 22(2): pp. 449–460.

56 Reicher, S.D., R. Spears, and T. Postmes, 'A social identity model of deindividuation phenomena', *European Review of Social Psychology*, 1995, 6(1): pp. 161–198.

57 Kanske, P., et al., 'Are strong empathizers better mentalizers? Evidence for independence and interaction between the routes of social cognition', *Social Cognitive and Affective Neuroscience*, 2016, 11(9): pp. 1383–1392.

58 Scherer, K.R., 'Appraisal theory', in *Handbook of Cognition and Emotion*, T. Dalgleish and M.J. Power (eds) (John Wiley & Sons, 1999), pp. 637–663.

59 Siemer, M., I. Mauss, and J.J. Gross, 'Same situation – different emotions: how appraisals shape our emotions', *Emotion*, 2007, 7(3): pp. 592–600.

60 Cherniss, C., 'Social and emotional competence in the workplace', in *The Handbook of Emotional Intelligence: Theory, Development, Assessment, and Application at Home, School, and in the Workplace*, R. Bar-On and J.D.A. Parker (eds) (Jossey-Bass, 2000), pp. 433–458.

61 Dewe, P., 'Primary appraisal, secondary appraisal and coping: their role in stressful work encounters', *Journal of Occupational Psychology*, 1991, 64(4): pp. 331–351.

62 Kalter, J., 'The workplace burnout', *Columbia Journalism Review*, 1999, 38(2): p. 30.

63 Zapf, D., et al., 'Emotion work and job stressors and their effects on burnout', *Psychology & Health*, 2001, 16(5): pp. 527–545.

64 Biegler, P., 'Autonomy, stress, and treatment of depression', *BMJ*, 2008, 336(7652): pp. 1046–1048.

65 Willner, P., et al., 'Loss of social status: preliminary evaluation of a novel animal model of depression', *Journal of Psychopharmacology*, 1995, 9(3): pp. 207–213.

66 Siegrist, J., et al., 'A short generic measure of work stress in the era of globalization: effort–reward imbalance', *International Archives of Occupational and Environmental Health*, 2009, 82(8): p. 1005.

67   Norris, C.J., et al., 'The interaction of social and emotional processes in the brain', *Journal of Cognitive Neuroscience*, 2004, 16(10): pp. 1818–1829.

68   Joyce, S., et al., 'Road to resilience: a systematic review and meta-analysis of resilience training programmes and interventions', *BMJ Open*, 2018, 8(6).

69   Thummakul, D., et al. (2012), 'The development of happy workplace index', *International Journal of Business Management*, 2012, 1(2): pp. 527–536.

70   Mann, A. and J. Harter, 'The worldwide employee engagement crisis', *Gallup Business Journal*, 2016, 7: pp. 1–5.

71   Hosie, P. and N. ElRakhawy, 'The happy worker: revisiting the "happy–productive worker" thesis', in *Wellbeing: A Complete Reference Guide*, Vol. 3, P.Y. Chen and C.L. Cooper (eds) (Wiley-Blackwell, 2014): pp. 113–138.

72   Miron, A.M. and J.W. Brehm, 'Reactance theory – 40 years later', *Zeitschrift für Sozialpsychologie*, 2006, 37(1): pp. 9–18.

73   Wagner, D.T., C.M. Barnes, and B.A. Scott, 'Driving it home: how workplace emotional labor harms employee home life', *Personnel Psychology*, 2014, 67(2): pp. 487–516.

74   Impett, E.A., et al., 'Suppression sours sacrifice: emotional and relational costs of suppressing emotions in romantic relationships', *Personality and Social Psychology Bulletin*, 2012, 38(6): pp. 707–720.

75   Flynn, J.J., T. Hollenstein, and A. Mackey, 'The effect of suppressing and not accepting emotions on depressive symptoms: is suppression different for men and women?' *Personality and Individual Differences*, 2010, 49(6): pp. 582–586.

76   Yoon, J.-H., et al., 'Suppressing emotion and engaging with complaining customers at work related to experience of depression and anxiety symptoms: a nationwide cross-sectional study', *Industrial Health*, 2017, 55: pp. 265–274.

77   Taylor, L., 'Out of character: how acting puts a mental strain on performers', *The Conversation*, 6 December 2017.

78   Durand, F., C. Isaac, and D. Januel, 'Emotional memory in post-traumatic stress disorder: a systematic PRISMA review of controlled studies', *Frontiers in Psychology*, 2019, 10(303).

79   Maxwell, I., M. Seton, and M. Szabó, 'The Australian actors' wellbeing study: a preliminary report', *About Performance*, 2015, 13: pp. 69–113.

80   Arias, G.L., 'In the wings: actors & mental health a critical review of the literature', Masters thesis, 2019, Lesley University.

81   Taylor, 'Out of character'.

82   Jones, P., *Drama as Therapy: Theatre as Living* (Psychology Press, 1996).

83   Cerney, M.S. and J.R. Buskirk, 'Anger: the hidden part of grief', *Bulletin of the Menninger Clinic*, 1991, 55(2): p. 228.

84 McCracken, L.M., 'Anger, injustice, and the continuing search for psychological mechanisms of pain, suffering, and disability', *Pain*, 2013, 154(9): pp. 1495–1496.

85 Kübler-Ross, E. and D. Kessler, *On Grief and Grieving: Finding the Meaning of Grief Through the Five Stages of Loss* (Simon and Schuster, 2005).

86 Silani, G., et al., 'Right supramarginal gyrus is crucial to overcome emotional egocentricity bias in social judgments', *Journal of Neuroscience*, 2013, 33(39): pp. 15466–15476.

87 Lamm, C., M. Rütgen, and I.C. Wagner, 'Imaging empathy and prosocial emotions', *Neuroscience Letters*, 2019, 693: pp. 49–53.

88 Carlson, N.R., *Physiology of Behavior* (Pearson Higher Education, 2012).

89 Silani, et al., 'Right supramarginal gyrus is crucial'.

90 Chang, S.W., et al., 'Neural mechanisms of social decision-making in the primate amygdala', *Proceedings of the National Academy of Sciences*, 2015, 112(52): pp. 16012–16017.

91 Stietz, et al., 'Dissociating empathy from perspective-taking'.

92 Hein, G. and R.T. Knight, 'Superior temporal sulcus – it's my area: or is it?' *Journal of Cognitive Neuroscience*, 2008, 20(12): pp. 2125–2136.

93 Dvash, J. and S.G. Shamay-Tsoory, 'Theory of mind and empathy as multidimensional constructs: neurological foundations', *Topics in Language Disorders*, 2014, 34(4): pp. 282–295.

94 Joireman, J.A., T.L. Needham, and A.-L. Cummings, 'Relationships between dimensions of attachment and empathy', *North American Journal of Psychology*, 2002, 4(1): pp. 63–80.

95 Hall, J.A. and S.E. Taylor, 'When love is blind: maintaining idealized images of one's spouse', *Human Relations*, 1976, 29(8): pp. 751–761.

96 Milton, D.E., 'On the ontological status of autism: the "double empathy problem"', *Disability & Society*, 2012, 27(6): pp. 883–887.

97 De Waal, 'Putting the altruism back into altruism'.

98 Cikara, M., et al., 'Their pain gives us pleasure: how intergroup dynamics shape empathic failures and counter-empathic responses', *Journal of Experimental Social Psychology*, 2014, 55: pp. 110–125.

99 Cikara, M., E.G. Bruneau, and R.R. Saxe, 'Us and them: intergroup failures of empathy', *Current Directions in Psychological Science*, 2011, 20(3): pp. 149–153.

100 Pezdek, K., I. Blandon-Gitlin, and C. Moore, 'Children's face recognition memory: more evidence for the cross-race effect', *Journal of Applied Psychology*, 2003, 88(4): p. 760.

101   Chiao, J.Y. and V.A. Mathur, 'Intergroup empathy: how does race affect empathic neural responses?', *Current Biology*, 2010, 20(11): pp. R478–R480.

102   Riess, 'The science of empathy'.

103   Stevens, F.L. and A.D. Abernethy, 'Neuroscience and racism: the power of groups for overcoming implicit bias', *International Journal of Group Psychotherapy*, 2018, 68(4): pp. 561–584.

104   Reyes, B.N., S.C. Segal, and M.C. Moulson, 'An investigation of the effect of race-based social categorization on adults' recognition of emotion', *PLOS One*, 2018, 13(2): p. e0192418.

105   Cikara, M. and S.T. Fiske, 'Bounded empathy: neural responses to outgroup targets' (mis)fortunes', *Journal of Cognitive Neuroscience*, 2011, 23(12): pp. 3791–3803.

106   Tadmor, C.T., et al., 'Multicultural experiences reduce intergroup bias through epistemic unfreezing', *Journal of Personality and Social Psychology*, 2012, 103(5): p. 750.

107   Riess, 'The science of empathy'.

108   General Medical Council, *Personal Beliefs and Medical Practice* (General Medical Council, 2008).

109   Doulougeri, K., E. Panagopoulou, and A. Montgomery, '(How) do medical students regulate their emotions?' *BMC Medical Education*, 2016, 16(1): p. 312.

110   Boissy, A., et al., 'Communication skills training for physicians improves patient satisfaction', *Journal of General Internal Medicine*, 2016, 31(7): pp. 755–761.

111   Flannelly, K.J., et al., 'The correlates of chaplains' effectiveness in meeting the spiritual/religious and emotional needs of patients', *Journal of Pastoral Care & Counseling*, 2009, 63(1–2): pp. 1–16.

112   Morgan, M., *Critical: Stories from the Front Line of Intensive Care Medicine* (Simon and Schuster, 2019).

113   Cameron, C., *Resolving Childhood Trauma: A Long-term Study of Abuse Survivors* (Sage, 2000).

5. Emotional Relationships

1   Batson, C.D., et al., 'An additional antecedent of empathic concern: valuing the welfare of the person in need', *Journal of Personality and Social Psychology*, 2007, 93(1): p. 65.

2   John, O.P. and J.J. Gross, 'Healthy and unhealthy emotion regulation: personality processes, individual differences, and life span development', *Journal of Personality*, 2004, 72(6): pp. 1301–1334.

3   O'Higgins, M., et al., 'Mother-child bonding at 1 year; associations with symptoms of postnatal depression and bonding in the first few weeks', *Archives of Women's Mental Health*, 2013, 16(5): pp. 381–389.

4   Wee, K.Y., et al., 'Correlates of ante- and postnatal depression in fathers: a systematic review', *Journal of Affective Disorders*, 2011, 130(3): pp. 358–377.

5   Althammer, F. and V. Grinevich, 'Diversity of oxytocin neurones: beyond magno-and parvocellular cell types?' *Journal of Neuroendocrinology*, 2018, 30(8): p. e12549.

6   Schneiderman, I., et al., 'Oxytocin during the initial stages of romantic attachment: relations to couples' interactive reciprocity', *Psychoneuroendocrinology*, 2012, 37(8): pp. 1277–1285.

7   Gravotta, L., 'Be mine forever: oxytocin may help build long-lasting love', *Scientific American*, 12 February 2013.

8   Magon, N. and S. Kalra, 'The orgasmic history of oxytocin: love, lust, and labor', *Indian Journal of Endocrinology and Metabolism*, 2011, 15(7): p. 156.

9   Scheele, D., et al., 'Oxytocin modulates social distance between males and females', *The Journal of Neuroscience*, 2012, 32(46): pp. 16074–16079.

10   Scheele, D., et al., 'Oxytocin enhances brain reward system responses in men viewing the face of their female partner', *Proceedings of the National Academy of Sciences*, 2013, 110(50): pp. 20308–20313.

11   Fineberg, S.K. and D.A. Ross, 'Oxytocin and the social brain', *Biological Psychiatry*, 2017, 81(3): p. e19.

12   Ross, H.E. and L.J. Young, 'Oxytocin and the neural mechanisms regulating social cognition and affiliative behavior', *Frontiers in Neuroendocrinology*, 2009, 30(4): pp. 534–547.

13   Guastella, A.J., P.B. Mitchell, and F. Mathews, 'Oxytocin enhances the encoding of positive social memories in humans', *Biological Psychiatry*, 2008, 64(3): pp. 256–258.

14   Bartz, J.A., et al., 'Social effects of oxytocin in humans: context and person matter', *Trends in Cognitive Sciences*, 2011, 15(7): pp. 301–309.

15   Shamay-Tsoory, S.G., et al., 'Intranasal administration of oxytocin increases envy and schadenfreude (gloating)', *Biological Psychiatry*, 2009, 66(9): pp. 864–870.

16   De Dreu, C.K.W., et al., 'Oxytocin promotes human ethnocentrism', *Proceedings of the National Academy of Sciences*, 2011, 108(4): pp. 1262–1266.

17   Flinn, M.V., D.C. Geary, and C.V. Ward, 'Ecological dominance, social competition, and coalitionary arms races: why humans evolved extraordinary intelligence', *Evolution and Human Behavior*, 2005, 26(1): pp. 10–46.

18 Nephew, B.C., 'Behavioral roles of oxytocin and vasopressin', in *Neuroendocrinology and Behavior*, T. Sumiyoshi (ed.) (InTech, 2012).

19 Bales, K.L., et al., 'Neural correlates of pair-bonding in a monogamous primate', *Brain Research*, 2007, 1184: pp. 245–253.

20 Knobloch, H. and V. Grinevich, 'Evolution of oxytocin pathways in the brain of vertebrates', *Frontiers in Behavioral Neuroscience*, 2014, 8(31).

21 Gruber C. W., 'Physiology of invertebrate oxytocin and vasopressin neuropeptides, *Experimental Physiology*, 2014, 99(1): pp. 55–61.

22 Nissen, E., et al., 'Elevation of oxytocin levels early post partum in women', *Acta Obstetricia et Gynecologica Scandinavica*, 1995, 74(7): pp. 530–533.

23 Ross and Young, 'Oxytocin and the neural mechanisms'.

24 Buckley, S.J., 'Ecstatic birth: the hormonal blueprint of labor', *Mothering Magazine*, 2002, 111: pp. 59–68.

25 Moberg, K.U. and D.K. Prime, 'Oxytocin effects in mothers and infants during breastfeeding', *Infant*, 2013, 9(6): pp. 201–206.

26 Wan, M.W., et al., 'The neural basis of maternal bonding', *PLOS One*, 2014, 9(3): p. e88436.

27 Leknes, S., et al., 'Oxytocin enhances pupil dilation and sensitivity to "hidden" emotional expressions', *Social Cognitive and Affective Neuroscience*, 2013, 8(7): pp. 741–749.

28 Vittner, D., et al., 'Increase in oxytocin from skin-to-skin contact enhances development of parent–infant relationship', *Biological Research for Nursing*, 2018, 20(1): pp. 54–62.

29 Peterman, K., 'What's love got to do with it? The potential role of oxytocin in the association between postpartum depression and mother-to-infant skin-to-skin contact', Masters thesis, 2014, University of North Carolina at Chapel Hill.

30 Young, K.S., et al., 'The neural basis of responsive caregiving behaviour: investigating temporal dynamics within the parental brain', *Behavioural Brain Research*, 2017, 325: pp. 105–116.

31 Glocker, M.L., et al., 'Baby schema in infant faces induces cuteness perception and motivation for caretaking in adults', *Ethology: Formerly Zeitschrift für Tierpsychologie*, 2009, 115(3): pp. 257–263.

32 Moberg and Prime, 'Oxytocin effects in mothers and infants'.

33 Peltola, M.J., L. Strathearn, and K. Puura, 'Oxytocin promotes face-sensitive neural responses to infant and adult faces in mothers', *Psychoneuroendocrinology*, 2018, 91: pp. 261–270.

34   Stavropoulos, K.K.M. and L.A. Alba, "'It's so cute I could crush it!'":
     understanding neural mechanisms of cute aggression', *Frontiers in Behavioral Neuroscience*, 2018, 12(300).

35   Kuzawa, C.W., et al., 'Metabolic costs and evolutionary implications of
     human brain development', *Proceedings of the National Academy of Sciences*,
     2014, 111(36): pp. 13010–13015.

36   Borgi, M., et al., 'Baby schema in human and animal faces induces cuteness
     perception and gaze allocation in children', *Frontiers in Psychology*, 2014, 5:
     p. 411.

37   Kringelbach, M.L., et al., 'On cuteness: unlocking the parental brain and
     beyond', *Trends in Cognitive Sciences*, 2016, 20(7): pp. 545–558.

38   Stavropoulos and Alba, "'It's so cute I could crush it!'".

39   Stavropoulos and Alba, "'It's so cute I could crush it!'".

40   Carter, C.S., 'The oxytocin–vasopressin pathway in the context of love and
     fear', *Frontiers in Endocrinology*, 2017, 8: p. 356.

41   Carter, C.S., 'Oxytocin pathways and the evolution of human behavior',
     Annual Review of Psychology, 2014, 65: pp. 17–39.

42   Bosch, O.J. and I.D. Neumann, 'Vasopressin released within the central
     amygdala promotes maternal aggression', *European Journal of Neuroscience*,
     2010, 31(5): pp. 883–891.

43   Carter, 'The oxytocin–vasopressin pathway'.

44   Sullivan, R., et al., 'Infant bonding and attachment to the caregiver: insights
     from basic and clinical science', *Clinics in Perinatology*, 2011, 38(4):
     pp. 643–655.

45   Choi, C.Q., 'Juvenile thoughts', *Scientific American*, 2009, 301(1): pp. 23–24.

46   Lukas, M., et al., 'The neuropeptide oxytocin facilitates pro-social behavior
     and prevents social avoidance in rats and mice', *Neuropsychopharmacology*,
     2011, 36(11): pp. 2159–2168.

47   Tomasello, M., 'The ultra-social animal', *European Journal of Social Psychology*,
     2014, 44(3): pp. 187–194.

48   Carter, C.S., 'The role of oxytocin and vasopressin in attachment',
     Psychodynamic Psychiatry, 2017, 45(4): pp. 499–517.

49   Carter, 'Oxytocin pathways and the evolution of human behavior'.

50   Morman, M.T. and K. Floyd, 'A "changing culture of fatherhood": effects on
     affectionate communication, closeness, and satisfaction in men's relationships
     with their fathers and their sons', *Western Journal of Communication (includes
     Communication Reports)*, 2002, 66(4): pp. 395–411.

51   Moir, A. and D. Jessel, *Brain Sex* (Random House, 1997).

52 DeLamater, J. and W.N. Friedrich, 'Human sexual development', *Journal of Sex Research*, 2002, 39(1): pp. 10–14.

53 Rippon, G., *The Gendered Brain: The New Neuroscience that Shatters the Myth of the Female Brain* (Random House, 2019).

54 Simmons, J.G., *The Scientific 100: A Ranking of the Most Influential Scientists, Past and Present* (Citadel Press, 2000).

55 Valine, Y.A., 'Why cultures fail: the power and risk of Groupthink', *Journal of Risk Management in Financial Institutions*, 2018, 11(4): pp. 301–307.

56 Simmons, *The Scientific 100*.

57 Bergman, G., 'The history of the human female inferiority ideas in evolutionary biology', *Rivista di Biologia*, 2002, 95(3): pp. 379–412.

58 Krulwich, R., 'Non! Nein! No! A country that wouldn't let women vote till 1971', *National Geographic*, 26 August 2016.

59 Clarke, E.H., *Sex in Education, Or, A Fair Chance for Girls* (James R. Osgood and Company, 1874).

60 Thompson, L., *The Wandering Womb: A Cultural History of Outrageous Beliefs about Women* (Prometheus Books, 2012)

61 Milne-Smith, A., 'Hysterical men: the hidden history of male nervous illness', *Canadian Journal of History*, 2009, 44(2): p. 365.

62 Tierney, A.J., 'Egas Moniz and the origins of psychosurgery: a review commemorating the 50th anniversary of Moniz's Nobel Prize', *Journal of the History of the Neurosciences*, 2000, 9(1): pp. 22–36.

63 Tone, A. and M. Koziol, '(F)ailing women in psychiatry: lessons from a painful past', *Canadian Medical Association Journal (CMAJ)*, 2018, 190(20): pp. E624–E625.

64 Tone and Koziol, '(F)ailing women in psychiatry'.

65 Baron-Cohen, S., 'The extreme male brain theory of autism', *Trends in Cognitive Sciences*, 2002, 6(6): pp. 248–254.

66 Lawson, J., S. Baron-Cohen, and S. Wheelwright, 'Empathising and systemising in adults with and without Asperger syndrome', *Journal of Autism and Developmental Disorders*, 2004, 34(3): pp. 301–310.

67 Andrew, J., M. Cooke, and S. Muncer, 'The relationship between empathy and Machiavellianism: an alternative to empathizing–systemizing theory', *Personality and Individual Differences*, 2008, 44(5): pp. 1203–1211.

68 Baez, S., et al., 'Men, women . . . who cares? A population-based study on sex differences and gender roles in empathy and moral cognition', *PLOS One*, 2017, 12(6): p. e0179336.

69 Ridley, R., 'Some difficulties behind the concept of the 'Extreme male brain' in autism research. A theoretical review', *Research in Autism Spectrum Disorders*, 2019, 57: pp. 19–27.

70 Gould, J. and J. Ashton-Smith, 'Missed diagnosis or misdiagnosis? Girls and women on the autism spectrum', *Good Autism Practice (GAP)*, 2011, 12(1): pp. 34–41.

71 Peters, M., 'Sex differences in human brain size and the general meaning of differences in brain size', *Canadian Journal of Psychology/Revue canadienne de psychologie*, 1991, 45(4): p. 507.

72 Rushton, J.P. and C.D. Ankney, 'Whole brain size and general mental ability: a review', *International Journal of Neuroscience*, 2009, 119(5): pp. 692–732.

73 Luders, E. and F. Kurth, 'Structural differences between male and female brains', in *Handbook of Clinical Neurology* (Elsevier, 2020), pp. 3–11.

74 Seifritz, E., et al., 'Differential sex-independent amygdala response to infant crying and laughing in parents versus nonparents', *Biological Psychiatry*, 2003, 54(12): pp. 1367–1375.

75 Stevens, F.L., R.A. Hurley, and K.H. Taber, 'Anterior cingulate cortex: unique role in cognition and emotion', *The Journal of Neuropsychiatry and Clinical Neurosciences*, 2011, 23(2): pp. 121–125.

76 Kong, F., et al., 'Sex-related neuroanatomical basis of emotion regulation ability', *PLOS One*, 2014, 9(5): p. e97071.

77 Stevens, J.S. and S. Hamann, 'Sex differences in brain activation to emotional stimuli: a meta-analysis of neuroimaging studies', *Neuropsychologia*, 2012, 50(7): pp. 1578–1593.

78 Wharton, W., et al., 'Neurobiological underpinnings of the estrogen-mood relationship', *Current Psychiatry Reviews*, 2012, 8(3): pp. 247–256.

79 McCarthy, M., 'Estrogen modulation of oxytocin and its relation to behavior', *Advances in Experimental Medicine and Biology*, 1995, 395: pp. 235–245.

80 Votinov, M., et al., 'Effects of exogenous testosterone application on network connectivity within emotion regulation systems', *Scientific Reports*, 2020, 10(1): pp. 1–10.

81 Baez, et al., 'Men, women . . . who cares?'.

82 Minor, M.W., 'Experimenter-expectancy effect as a function of evaluation apprehension', *Journal of Personality and Social Psychology*, 1970, 15(4): p. 326.

83 Dreher, J.-C., et al., 'Testosterone causes both prosocial and antisocial status-enhancing behaviors in human males', *Proceedings of the National Academy of Sciences*, 2016, 113(41): pp. 11633–11638.

84   Sapolsky, R.M., 'Doubled-edged swords in the biology of conflict', *Frontiers in Psychology*, 2018, 9: p. 2625.

85   Zink, C.F., et al., 'Know your place: neural processing of social hierarchy in humans', *Neuron*, 2008, 58(2): pp. 273–283.

86   Tabibnia, G. and M.D. Lieberman, 'Fairness and cooperation are rewarding', *Annals of the New York Academy of Sciences*, 2007, 1118(1): pp. 90–101.

87   Tabibnia and Lieberman, 'Fairness and cooperation are rewarding'.

88   Eisenegger, C., et al., 'Prejudice and truth about the effect of testosterone on human bargaining behaviour', *Nature*, 2010, 463(7279): pp. 356–359.

89   Wibral, M., et al., 'Testosterone administration reduces lying in men', *PLOS One*, 2012, 7(10): p. e46774.

90   Maguire, E.A., K. Woollett, and H.J. Spiers, 'London taxi drivers and bus drivers: a structural MRI and neuropsychological analysis', *Hippocampus*, 2006, 16(12): pp. 1091–1101.

91   Kaplow, J.B., et al., 'Emotional suppression mediates the relation between adverse life events and adolescent suicide: implications for prevention', *Prevention Science*, 2014, 15(2): pp. 177–185.

92   Albert, P.R., 'Why is depression more prevalent in women?' *Journal of Psychiatry & Neuroscience: JPN*, 2015, 40(4): p. 219.

93   Hedegaard, H., S.C. Curtin, and M. Warner, 'Suicide rates in the United States continue to increase', *NCHS Data Brief*, 2018, 309.

94   Noone, P.A., 'The Holmes–Rahe Stress Inventory', *Occupational Medicine*, 2017, 67(7): pp. 581–582.

95   Kim, J. and E. Hatfield, 'Love types and subjective well-being: a cross-cultural study', *Social Behavior and Personality: An International Journal*, 2004, 32(2): pp. 173–182.

96   Lewis, M., J.M. Haviland-Jones, and L.F. Barrett, *Handbook of Emotions* (Guilford Press, 2010).

97   Cacioppo, S., et al., 'Social neuroscience of love', *Clinical Neuropsychiatry*, 2012, 9(1), pp. 3–13.

98   Barsade, S.G. and O.A. O'Neill, 'What's love got to do with it? A longitudinal study of the culture of companionate love and employee and client outcomes in a long-term care setting', *Administrative Science Quarterly*, 2014, 59(4): pp. 551–598.

99   Gilbert, D.T., S.T. Fiske, and G. Lindzey, *The Handbook of Social Psychology*, Vol. 1 (Oxford University Press, 1998).

100   Bartels, A. and S. Zeki, 'The neural correlates of maternal and romantic love', *NeuroImage*, 2004, 21(3): pp. 1155–1166.

101   Ainsworth, M.D.S., et al., *Patterns of Attachment: A Psychological Study of the Strange Situation* (Psychology Press, 2015).

102   Purves, D., G. Augustine, and D. Fitzpatrick, *Autonomic Regulation of Sexual Function* (Sinauer Associates, 2001).

103   Benson, E. 'The science of sexual arousal', 2003. Available from: http://www.apa.org/monitor/apr03/arousal.aspx.

104   Herzberg, L.A., 'On sexual lust as an emotion', *HUMANA.MENTE Journal of Philosophical Studies*, 2019, 12(35): pp. 271–302.

105   Bogaert, A.F., 'Asexuality: what it is and why it matters', *Journal of Sex Research*, 2015, 52(4): pp. 362–379.

106   Chasin, C.D., 'Making sense in and of the asexual community: navigating relationships and identities in a context of resistance', *Journal of Community & Applied Social Psychology*, 2015, 25(2): pp. 167–180.

107   Cacioppo, S., et al., 'The common neural bases between sexual desire and love: a multilevel kernel density fMRI analysis', *The Journal of Sexual Medicine*, 2012, 9(4): pp. 1048–1054.

108   Cacioppo, et al., 'The common neural bases'.

109   Takahashi, K., et al., 'Imaging the passionate stage of romantic love by dopamine dynamics', *Frontiers in Human Neuroscience*, 2015, 9: p. 191.

110   Volkow, N.D., G.-J. Wang, and R.D. Baler, 'Reward, dopamine and the control of food intake: implications for obesity', *Trends in Cognitive Sciences*, 2011, 15(1): pp. 37–46.

111   Villablanca, J.R., 'Why do we have a caudate nucleus?', *Acta Neurobiologiae Experimentalis (Wars)*, 2010, 70(1): pp. 95–105.

112   Ainsworth, et al., *Patterns of Attachment*.

113   Helmuth, L., 'Caudate-over-heels in love', *Science*, 2003, 302(5649): p. 1320.

114   Bartels and Zeki, 'The neural correlates of maternal and romantic love'.

115   Chowdhury, R., et al., 'Dopamine modulates episodic memory persistence in old age', *Journal of Neuroscience*, 2012, 32(41): pp. 14193–14204.

116   Raderschall, et al., 'Habituation under natural conditions'.

117   Fisher, H.E., et al., 'Reward, addiction, and emotion regulation systems associated with rejection in love', *Journal of Neurophysiology*, 2010, 104(1): pp. 51–60.

118   Myers Ernst, M. and L.H. Epstein, 'Habituation of responding for food in humans', *Appetite*, 2002, 38(3): pp. 224–234.

119   Acevedo, B.P. and A. Aron, 'Does a long-term relationship kill romantic love?' *Review of General Psychology*, 2009, 13(1): pp. 59–65.

120   Masuda, M., 'Meta-analyses of love scales: do various love scales measure the same psychological constructs?' *Japanese Psychological Research*, 2003, 45(1): pp. 25–37.

121   Horstman, A.M., et al., 'The role of androgens and estrogens on healthy aging and longevity', *Journals of Gerontology Series A: Biomedical Sciences and Medical Sciences*, 2012, 67(11): pp. 1140–1152.

122   Kılıç, N. and A. Altınok, 'Obsession and relationship satisfaction through the lens of jealousy and rumination', *Personality and Individual Differences*, 2021, 179: p. 110959.

123   Harris, C.R., 'Sexual and romantic jealousy in heterosexual and homosexual adults', *Psychological Science*, 2002, 13(1): pp. 7–12.

124   Richards, J.M., E.A. Butler, and J.J. Gross, 'Emotion regulation in romantic relationships: the cognitive consequences of concealing feelings', *Journal of Social and Personal Relationships*, 2003, 20(5): pp. 599–620.

125   Ellsworth, P.C., 'Appraisal theory: old and new questions', *Emotion Review*, 2013, 5(2): pp. 125–131.

126   Field, T., 'Romantic breakups, heartbreak and bereavement – romantic breakups', *Psychology*, 2011, 2(4): p. 382.

127   Davis, M.H. and H.A. Oathout, 'Maintenance of satisfaction in romantic relationships: empathy and relational competence', *Journal of Personality and Social Psychology*, 1987, 53(2): p. 397.

128   Acevedo and Aron, 'Does a long-term relationship kill romantic love?'.

129   Diener, E., et al., 'Subjective well-being: three decades of progress', *Psychological Bulletin*, 1999, 125(2): p. 276.

130   Aron, A., et al., 'Reward, motivation, and emotion systems associated with early-stage intense romantic love', *Journal of Neurophysiology*, 2005, 94(1): pp. 327–337.

131   Arzy, S., et al., 'Induction of an illusory shadow person', *Nature*, 2006, 443: p. 287.

132   Lamb, M.E. and C. Lewis, 'The role of parent–child relationships in child development', in *Social and Personality Development*, M.E. Lamb and M.H. Bornstein (eds) (Psychology Press, 2013), pp. 267–316.

133   Silverberg, S.B. and L. Steinberg, 'Adolescent autonomy, parent-adolescent conflict, and parental well-being', *Journal of Youth and Adolescence*, 1987, 16(3): pp. 293–312.

134   Aquilino, W.S., 'From adolescent to young adult: a prospective study of parent-child relations during the transition to adulthood', *Journal of Marriage and the Family*, 1997, 59(3): pp. 670–686.

135 Ro, C., 'Dunbar's number: why we can only maintain 150 relationships', BBC Future, accessed July 2020.

136 Lindenfors, P., A. Wartel, and J. Lind, '"Dunbar's number" deconstructed', *Biology Letters*, 2021, 17(5): p. 20210158.

137 Ro, 'Dunbar's number'.

138 Ampel, B.C., M. Muraven, and E.C. McNay, 'Mental work requires physical energy: self-control is neither exception nor exceptional', *Frontiers in Psychology*, 2018, 9: p. 1005.

139 Schwartz, B., 'The social psychology of privacy', *American Journal of Sociology*, 1968, 73(6): pp. 741–752.

140 Giles, D.C., 'Parasocial interaction: a review of the literature and a model for future research', *Media Psychology*, 2002, 4(3): pp. 279–305.

141 Schiappa, E., M. Allen, and P.B. Gregg, 'Parasocial relationships and television: a meta-analysis of the effects', in *Mass Media Effects Research: Advances Through Meta-analysis*, R.W. Preiss et al. (eds) (Routledge, 2007), pp. 301–314.

142 Allen, P., et al., 'The hallucinating brain: a review of structural and functional neuroimaging studies of hallucinations', *Neuroscience & Biobehavioral Reviews*, 2008, 32(1): pp. 175–191.

143 Blakemore, S.-J., et al., 'The perception of self-produced sensory stimuli in patients with auditory hallucinations and passivity experiences: evidence for a breakdown in self-monitoring', *Psychological Medicine*, 2000, 30(5): pp. 1131–1139.

144 Behrmann, M., 'The mind's eye mapped onto the brain's matter', *Current Directions in Psychological Science*, 2000, 9(2): pp. 50–54.

145 Mullally, S.L. and E.A. Maguire, 'Memory, imagination, and predicting the future: a common brain mechanism?' *The Neuroscientist*, 2014, 20(3): pp. 220–234.

146 Hemmer and Steyvers, 'A Bayesian account'.

147 Buckner, R.L., 'The role of the hippocampus in prediction and imagination', *Annual Review of Psychology*, 2010, 61: pp. 27–48.

148 Hassabis, D. and E.A. Maguire, 'Deconstructing episodic memory with construction', *Trends in Cognitive Sciences*, 2007, 11(7): pp. 299–306.

149 Spreng, R.N., R.A. Mar, and A.S. Kim, 'The common neural basis of autobiographical memory, prospection, navigation, theory of mind, and the default mode: a quantitative meta-analysis', *Journal of Cognitive Neuroscience*, 2009, 21(3): pp. 489–510.

150   Diekhof, E.K., et al., 'The power of imagination – how anticipatory mental imagery alters perceptual processing of fearful facial expressions', *NeuroImage*, 2011, 54(2): pp. 1703–1714.

151   Herz and von Clef, 'The influence of verbal labeling'.

152   Henderson, R.R., M.M. Bradley, and P.J. Lang, 'Emotional imagery and pupil diameter', *Psychophysiology*, 2018, 55(6): p. e13050.

153   Perse, E.M. and R.B. Rubin, 'Attribution in social and parasocial relationships', *Communication Research*, 1989, 16(1): pp. 59–77.

154   Brown, W.J., 'Examining four processes of audience involvement with media personae: transportation, parasocial interaction, identification, and worship', *Communication Theory*, 2015, 25(3): pp. 259–283.

155   Hineline, P.N., 'Narrative: why it's important, and how it works', *Perspectives on Behavior Science*, 2018, 41(2): pp. 471–501.

156   Green, M.C., 'Transportation into narrative worlds: the role of prior knowledge and perceived realism', *Discourse Processes*, 2004, 38(2): pp. 247–266.

157   Kelman, H., 'Processes of opinion change', *Public Opinion Quarterly*, 1961, 25: pp. 57–78.

158   Jenner, G., *Dead Famous: An Unexpected History of Celebrity from Bronze Age to Silver Screen* (Hachette, 2020).

159   Cohen, J., 'Defining identification: a theoretical look at the identification of audiences with media characters', *Mass Communication & Society*, 2001, 4(3): pp. 245–264.

160   Moyer-Gusé, E., A.H. Chung, and P. Jain, 'Identification with characters and discussion of taboo topics after exposure to an entertainment narrative about sexual health', *Journal of Communication*, 2011, 61(3): pp. 387–406.

161   Howard Gola, A.A., et al., 'Building meaningful parasocial relationships between toddlers and media characters to teach early mathematical skills', *Media Psychology*, 2013, 16(4): pp. 390–411.

162   Calvert, S.L., M.N. Richards, and C.C. Kent, 'Personalized interactive characters for toddlers' learning of seriation from a video presentation', *Journal of Applied Developmental Psychology*, 2014, 35(3): pp. 148–155.

163   Holt-Lunstad, J., 'The potential public health relevance of social isolation and loneliness: prevalence, epidemiology, and risk factors', *Public Policy & Aging Report*, 2017, 27(4): pp. 127–130.

164   Derrick, J.L., S. Gabriel, and B. Tippin, 'Parasocial relationships and self-discrepancies: faux relationships have benefits for low self-esteem individuals', *Personal Relationships*, 2008, 15(2): pp. 261–280.

165   Singer, J.L., 'Imaginative play and adaptive development', in *Toys, Play, and Child Development*, J.H. Goldstein (ed.) (Cambridge University Press, 1994), pp. 6–26.

166   Hoff, E.V., 'A friend living inside me – the forms and functions of imaginary companions', *Imagination, Cognition and Personality*, 2004, 24(2): pp. 151–189.

167   Taylor, M. and S.M. Carlson, 'The relation between individual differences in fantasy and theory of mind', *Child Development*, 1997, 68(3): pp. 436–455.

168   Pickhardt, C., 'Adolescence and the teenage crush', *Psychology Today*, 10 September 2012.

169   Erickson, S.E. and S. Dal Cin, 'Romantic parasocial attachments and the development of romantic scripts, schemas and beliefs among adolescents', *Media Psychology*, 2018, 21(1): pp. 111–136.

170   Knox, J., 'Sex, shame and the transcendent function: the function of fantasy in self development', *Journal of Analytical Psychology*, 2005, 50(5): pp. 617–639.

171   Tukachinksy, R., 'When actors don't walk the talk: parasocial relationships moderate the effect of actor-character incongruence', *International Journal of Communication*, 2015, 9: p. 17.

172   Proctor, W., '"Bitches ain't gonna hunt no ghosts": totemic nostalgia, toxic fandom and the *Ghostbusters* platonic', *Palabra Clave*, 2017, 20(4): pp. 1105–1141.

173   Biegler, 'Autonomy, stress'.

174   McCutcheon, L.E., et al., 'Exploring the link between attachment and the inclination to obsess about or stalk celebrities', *North American Journal of Psychology*, 2006, 8(2): pp. 289–300.

175   Pickhardt, 'Adolescence and the teenage crush'.

176   Eyal, K. and J. Cohen, 'When good friends say goodbye: a parasocial breakup study', *Journal of Broadcasting & Electronic Media*, 2006, 50(3): pp. 502–523.

6: Emotional Technology

1   Öhman, C.J. and D. Watson, 'Are the dead taking over Facebook? A Big Data approach to the future of death online', *Big Data & Society*, 2019, 6(1).

2   Kawamichi, et al., 'Increased frequency of social interaction'.

3   Krebs, et al., 'Novelty increases the mesolimbic functional connectivity'.

4   Farrow, T., et al., 'Neural correlates of self-deception and impression-management', *Neuropsychologia*, 2015, 67: pp. 159–174.

5   Dunbar, R. and R.I.M. Dunbar, *Grooming, Gossip, and the Evolution of Language* (Harvard University Press, 1998).

6   Dumas, G., et al., 'Inter-brain synchronization during social interaction', *PLOS One*, 2010, 5(8): p. e12166.

7   Van Baaren, et al., 'Where is the love?'.

8   Blanchard, et al., 'Risk assessment'.

9   Windeler, J.B., K.M. Chudoba, and R.Z. Sundrup, 'Getting away from them all: managing exhaustion from social interaction with telework', *Journal of Organizational Behavior*, 2017, 38(7): pp. 977–995.

10   Ross, S.A., 'Compensation, incentives, and the duality of risk aversion and riskiness', *The Journal of Finance*, 2004, 59(1): pp. 207–225.

11   Van Dillen, L.F. and H. van Steenbergen, 'Tuning down the hedonic brain: cognitive load reduces neural responses to high-calorie food pictures in the nucleus accumbens', *Cognitive, Affective, & Behavioral Neuroscience*, 2018, 18(3): pp. 447–459.

12   Legault and Inzlicht, 'Self-determination'.

13   Landhäußer, A. and J. Keller, 'Flow and its affective, cognitive, and performance-related consequences', in *Advances in Flow Research*, S. Engeser (ed.) (Springer, 2012), pp. 65–85.

14   Nakamura, J. and M. Csikszentmihalyi, 'The concept of flow', in *Flow and the Foundations of Positive Psychology: The Collected Works of Mihaly Csikszentmihalyi* (Springer, 2014), pp. 239–263.

15   Landhäußer and Keller, 'Flow'.

16   Nakamura and Csikszentmihalyi, 'The concept of flow'.

17   Sutcliffe, A.G., J.F. Binder, and R.I.M. Dunbar, 'Activity in social media and intimacy in social relationships', *Computers in Human Behavior*, 2018, 85: pp. 227–235.

18   Baltaci, Ö., 'The predictive relationships between the social media addiction and social anxiety, loneliness, and happiness', *International Journal of Progressive Education*, 2019, 15(4): pp. 73–82.

19   Buchholz, M., U. Ferm, and K. Holmgren, 'Support persons' views on remote communication and social media for people with communicative and cognitive disabilities', *Disability and Rehabilitation*, 2020, 42(10): pp. 1439–1447.

20   Hinduja, S. and J.W. Patchin, 'Cultivating youth resilience to prevent bullying and cyberbullying victimization', *Child Abuse & Neglect*, 2017, 73: pp. 51–62.

21   Whittaker, E. and R.M. Kowalski, 'Cyberbullying via social media', *Journal of School Violence*, 2015, 14(1): pp. 11–29.

22 Bottino, S.M.B., et al., 'Cyberbullying and adolescent mental health: systematic review', *Cadernos de Saude Publica*, 2015, 31: pp. 463–475.

23 Slonje, R. and P.K. Smith, 'Cyberbullying: another main type of bullying?' *Scandinavian Journal of Psychology*, 2008, 49(2): pp. 147–154.

24 Sticca, F. and S. Perren, 'Is cyberbullying worse than traditional bullying? Examining the differential roles of medium, publicity, and anonymity for the perceived severity of bullying', *Journal of Youth and Adolescence*, 2013, 42(5): pp. 739–750.

25 Tehrani, N., 'Bullying: a source of chronic post traumatic stress?' *British Journal of Guidance & Counselling*, 2004, 32(3): pp. 357–366.

26 Eisenberger, N.I., 'Why rejection hurts: what social neuroscience has revealed about the brain's response to social rejection', *Brain*, 2011, 3(2): p. 1.

27 Sticca and Perren, 'Is cyberbullying worse than traditional bullying?'.

28 Weiss, B. and R.S. Feldman, 'Looking good and lying to do it: deception as an impression management strategy in job interviews', *Journal of Applied Social Psychology*, 2006, 36(4): pp. 1070–1086.

29 Farrow, et al., 'Neural correlates of self-deception'.

30 Craven, R. and H.W. Marsh, 'The centrality of the self-concept construct for psychological wellbeing and unlocking human potential: implications for child and educational psychologists', *Educational & Child Psychology*, 2008, 25(2): pp. 104–118.

31 Akanbi, M.I. and A.B. Theophilus, 'Influence of social media usage on self-image and academic performance among senior secondary school students in Ilorin-West Local Goverment, Kwara State', *Research on Humanities and Social Sciences*, 2014, 4(14): pp. 58–62.

32 Tenney, E.R., et al., 'Calibration trumps confidence as a basis for witness credibility', *Psychological Science*, 2007, 18(1): pp. 46–50.

33 Bell, N.D., 'Responses to failed humor', *Journal of Pragmatics*, 2009, 41(9): pp. 1825–1836.

34 Emery, L.F., et al., 'Can you tell that I'm in a relationship? Attachment and relationship visibility on Facebook', *Personality and Social Psychology Bulletin*, 2014, 40(11): pp. 1466–1479.

35 Scott, K.M., et al., 'Associations between subjective social status and DSM-IV mental disorders: results from the World Mental Health surveys', *JAMA Psychiatry*, 2014, 71(12): pp. 1400–1408.

36 Kessler, R.C., 'Stress, social status, and psychological distress', *Journal of Health and Social Behavior*, 1979: pp. 259–272.

37 Verduyn, P., N. Gugushvili, and E. Kross, 'The impact of social network sites on mental health: distinguishing active from passive use', *World Psychiatry: Official Journal of the World Psychiatric Association (WPA)*, 2021, 20(1): pp. 133–134.

38 Escobar-Viera, C.G., et al., 'Passive and active social media use and depressive symptoms among United States adults', *Cyberpsychology, Behavior, and Social Networking*, 2018, 21(7): pp. 437–443.

39 Swist, T., et al., 'Social media and the wellbeing of children and young people: a literature review', 2015, Prepared for the Commissioner for Children and Young People, Western Australia.

40 Best, P., R. Manktelow, and B. Taylor, 'Online communication, social media and adolescent wellbeing: a systematic narrative review', *Children and Youth Services Review*, 2014, 41: pp. 27–36.

41 O'Reilly, M., et al., 'Is social media bad for mental health and wellbeing? Exploring the perspectives of adolescents', *Clinical Child Psychology and Psychiatry*, 2018, 23(4): pp. 601–613.

42 Burnett, S., et al., 'The social brain in adolescence: evidence from functional magnetic resonance imaging and behavioural studies', *Neuroscience & Biobehavioral Reviews*, 2011, 35(8): pp. 1654–1664.

43 Kleemans, M., et al., 'Picture perfect: the direct effect of manipulated Instagram photos on body image in adolescent girls', *Media Psychology*, 2018, 21(1): pp. 93–110.

44 O'Reilly, et al., 'Is social media bad?'.

45 Quinn, K., 'Social media and social wellbeing in later life', *Ageing & Society*, 2021, 41(6): pp. 1349–1370.

46 Gentner, D. and A.L. Stevens, *Mental Models* (Psychology Press, 2014).

47 Brehm, J.W. and A.R. Cohen, *Explorations in Cognitive Dissonance* (John Wiley & Sons, 1962).

48 Marris, P., *Loss and Change (Psychology Revivals): Revised Edition* (Routledge, 2014).

49 Hertenstein, M.J., et al., 'The communication of emotion via touch', *Emotion*, 2009, 9(4): p. 566.

50 Radulescu, A., 'Why do we walk around when talking on the phone?', *Medium*, 13 October 2020.

51 Oppezzo, M. and D.L. Schwartz, 'Give your ideas some legs: the positive effect of walking on creative thinking', *Journal of Experimental Psychology: Learning, Memory, and Cognition*, 2014, 40(4): p. 1142.

52 Lee, J., A. Jatowt, and K.S. Kim, 'Discovering underlying sensations of human emotions based on social media', *Journal of the Association for Information Science and Technology*, 2021, 72(4): pp. 417–432.

53 Gaither, S.E., et al., 'Thinking outside the box: multiple identity mind-sets affect creative problem solving', *Social Psychological and Personality Science*, 2015, 6(5): pp. 596–603.

54 Panger, G.T., *Emotion in Social Media* (UC Berkeley, 2017).

55 Hardicre, J., 'Valid informed consent in research: an introduction', *British Journal of Nursing*, 2014, 23(11): pp. 564–567.

56 Kramer, A.D.I., J.E. Guillory, and J.T. Hancock, 'Experimental evidence of massive-scale emotional contagion through social networks', *Proceedings of the National Academy of Sciences*, 2014, 111(24): pp. 8788–8790.

57 Goldenberg, A. and J.J. Gross, 'Digital emotion contagion', *Trends in Cognitive Sciences*, 2020, 24(4): pp. 316–328.

58 Burnett, G., M. Besant, and E.A. Chatman, 'Small worlds: normative behavior in virtual communities and feminist bookselling', *Journal of the Association for Information Science and Technology*, 2001, 52(7): p. 536.

59 Achar, C., et al., 'What we feel and why we buy: the influence of emotions on consumer decision-making', *Current Opinion in Psychology*, 2016, 10: pp. 166–170.

60 Utz, S., 'Social media as sources of emotions', in *Social Psychology in Action*, K. Sassenberg and M.L.W. Vliek (eds) (Springer, 2019), pp. 205–219.

61 Curtis, A., *The Power of Nightmares: The Rise of the Politics of Fear*, Documentary, BBC, 2004.

62 Ford, J.B., 'What do we know about celebrity endorsement in advertising?' *Journal of Advertising Research*, 2018, 58(1): pp. 1–2.

63 Bennet, J., 'The TSA is frighteningly awful at screening passengers', *Popular Mechanics*, 5 November 2015.

64 Anderson, N., 'TSA's got 94 signs to ID terrorists, but they're unproven by science', in *Ars Technica* (Condé Nast Digital, 2013).

65 Gendron, et al., 'Perceptions of emotion'.

66 Denault, V., et al., 'The analysis of nonverbal communication: the dangers of pseudoscience in security and justice contexts', *Anuario de Psicología Jurídica*, 2020, 30(1): pp. 1–12.

67 Butalia, M.A., M. Ingle, and P. Kulkarni, 'Facial expression recognition for security', *International Journal of Modern Engineering Research*, 2012, 2(4): pp. 1449–1453.

68 Wong, S.-L. and Q. Liu, 'Emotion recognition is China's new surveillance craze', *Financial Times*, 1 November 2019.

69 Matt, S.J., 'What the history of emotions can offer to psychologists, economists, and computer scientists (among others)', *History of Psychology*, 2021, 24(2): p. 121.

70 Ortmann, A. and R. Hertwig, 'The costs of deception: evidence from psychology', *Experimental Economics*, 2002, 5(2): pp. 111–131.

71 Warren, G., E. Schertler, and P. Bull, 'Detecting deception from emotional and unemotional cues', *Journal of Nonverbal Behavior*, 2009, 33(1): pp. 59–69.

72 Rodero, E. and I. Lucas, 'Synthetic versus human voices in audiobooks: the human emotional intimacy effect', *New Media & Society*, June 2021.

73 Liu, et al., 'Seeing Jesus in toast'.

74 Seyama, J. and R.S. Nagayama, 'The uncanny valley: effect of realism on the impression of artificial human faces', *Presence*, 2007, 16(4): pp. 337–351.

75 Lippmann, R.P., 'Neural nets for computing', *ICASSP*, 1988: pp. 1–6.

76 He, X. and W. Zhang, 'Emotion recognition by assisted learning with convolutional neural networks', *Neurocomputing*, 2018, 291: pp. 187–194.

77 Kornfield, R., et al., 'Detecting recovery problems just in time: application of automated linguistic analysis and supervised machine learning to an online substance abuse forum', *Journal of Medical Internet Research*, 2018, 20(6): p. e10136.

78 Birnbaum, M.L., et al., 'Detecting relapse in youth with psychotic disorders utilizing patient-generated and patient-contributed digital data from Facebook', *NPJ Schizophrenia*, 2019, 5(1): pp. 1–9.

79 Venkatapur, R.B., et al., 'THERABOT an artificial intelligent therapist at your fingertips', *IOSR Journal of Computer Engineering*, 2018, 20(3): pp. 34–38.

80 Craig, T.K., et al., 'AVATAR therapy for auditory verbal hallucinations in people with psychosis: a single-blind, randomised controlled trial', *The Lancet Psychiatry*, 2018, 5(1): pp. 31–40.

81 Kothgassner, O.D., et al., 'Virtual reality exposure therapy for posttraumatic stress disorder (PTSD): a meta-analysis', *European Journal of Psychotraumatology*, 2019, 10(1): p. 1654782.

82 Dunbar and Dunbar, *Grooming, Gossip*.

83 Wyse, D., *How Writing Works: From the Invention of the Alphabet to the Rise of Social Media* (Cambridge University Press, 2017).

84   Doosje, B.E., et al., 'Antecedents and consequences of group-based guilt: the effects of ingroup identification', *Group Processes & Intergroup Relations*, 2006, 9(3): pp. 325–338.

85   Lee, R.S., 'Credibility of newspaper and TV news', *Journalism Quarterly*, 1978, 55(2): pp. 282–287.

86   Jensen, J.D., et al., 'Public estimates of cancer frequency: cancer incidence perceptions mirror distorted media depictions', *Journal of Health Communication*, 2014, 19(5): pp. 609–624.

87   McCombs, M. and A. Reynolds, 'How the news shapes our civic agenda', in *Media Effects* (Routledge, 2009), pp. 17–32.

88   Desai, R.H., M. Reilly, and W. van Dam, 'The multifaceted abstract brain', *Philosophical Transactions of the Royal Society B: Biological Sciences*, 2018, 373(1752): p. 20170122.

89   Ampel, et al., 'Mental work requires physical energy'.

90   Cowan, N., 'The magical mystery four: how is working memory capacity limited, and why?', *Current Directions in Psychological Science*, 2010, 19(1): pp. 51–57.

91   Itti, L., 'Models of bottom-up attention and saliency', in *Neurobiology of Attention* (Elsevier, 2005), pp. 576–582.

92   Tyng, C.M., et al., 'The influences of emotion on learning and memory', *Frontiers in Psychology*, 2017, 8: p. 1454.

93   Howard Gola, et al., 'Building meaningful parasocial relationships'.

94   Zald and Pardo, 'Emotion, olfaction, and the human amygdala'.

95   Ungerer, F., 'Emotions and emotional language in English and German news stories', in *The Language of Emotions*, S. Niemeier and R. Dirven (eds) (John Benjamins, 1997), pp. 307–328.

96   Vlasceanu, M., J. Goebel, and A. Coman, 'The emotion-induced belief-amplification effect', *Proceedings of the 42nd Annual Conference of the Cognitive Science Society*, 2020: pp. 417–422.

97   Dreyer, K.J., et al., *A Guide to the Digital Revolution* (Springer, 2006).

98   de Melo, L.W.S., M.M. Passos, and R.F. Salvi, 'Analysis of 'flat-earther' posts on social media: reflections for science education from the discursive perspective of Foucault', *Revista Brasileira de Pesquisa em Educação em Ciências*, 2020, 20: pp. 295–313.

99   Dubois, E. and G. Blank, 'The echo chamber is overstated: the moderating effect of political interest and diverse media', *Information, Communication & Society*, 2018, 21(5): pp. 729–745.

100   Lowry, N. and D.W. Johnson, 'Effects of controversy on epistemic curiosity, achievement, and attitudes', *The Journal of Social Psychology*, 1981, 115(1): pp. 31–43.

101   Rozin and Royzman, 'Negativity bias'.

102   Trussler, M. and S. Soroka, 'Consumer demand for cynical and negative news frames', *The International Journal of Press/Politics*, 2014, 19(3): pp. 360–379.

103   Gorvett, Z., 'How the news changes the way we think and behave', BBC Future, 12 May 2020.

104   Asch, S.E., 'Studies of independence and conformity: I. A minority of one against a unanimous majority', *Psychological Monographs: General and Applied*, 1956, 70(9): p. 1.

105   Smaldino, P.E. and J.M. Epstein, 'Social conformity despite individual preferences for distinctiveness', *Royal Society Open Science*, 2015, 2(3): p. 140437.

106   Young, E., 'A new understanding: what makes people trust and rely on news', *American Press Institute*, April 2016.

107   Smith, T.B.M., 'Esoteric themes in David Icke's conspiracy theories', *Journal for the Academic Study of Religion*, 2017, 30(3): pp. 281–302.

108   Deutsch, M. and H.B. Gerard, 'A study of normative and informational social influences upon individual judgment', *The Journal of Abnormal and Social Psychology*, 1955, 51(3): p. 629.

109   Spanos, K.E., et al., 'Parent support for social media standards combatting vaccine misinformation', *Vaccine*, 2021, 39(9): pp. 1364–1369.

110   Wu, L., et al., 'Misinformation in social media: definition, manipulation, and detection', *ACM SIGKDD Explorations Newsletter*, 2019, 21(2): pp. 80–90.

111   Kenworthy, J.B., et al., 'Building trust in a postconflict society: an integrative model of cross-group friendship and intergroup emotions', *Journal of Conflict Resolution*, 2015, 60(6): pp. 1041–1070.

112   Mallinson, D.J. and P.K. Hatemi, 'The effects of information and social conformity on opinion change', *PLOS One*, 2018, 13(5): p. e0196600.

113   Cummins, R.G. and T. Chambers, 'How production value impacts perceived technical quality, credibility, and economic value of video news', *Journalism & Mass Communication Quarterly*, 2011, 88(4): pp. 737–752.

114   Abdulla, R.A., et al. 'The credibility of newspapers, television news, and online news', in *Education in Journalism Annual Convention, Florida USA* (Citeseer, 2002).

115 Tandoc Jr, E.C., 'Tell me who your sources are: perceptions of news credibility on social media', *Journalism Practice*, 2019, 13(2): pp. 178–190.

116 Wijenayake, S., et al., 'Effect of conformity on perceived trustworthiness of news in social media', *IEEE Internet Computing*, 2020, 25(1): pp. 12–19.

117 Janis, I.L., 'Groupthink', *IEEE Engineering Management Review*, 2008, 36(1): p. 36.

118 Lin, A., R. Adolphs, and A. Rangel, 'Social and monetary reward learning engage overlapping neural substrates', *Social Cognitive and Affective Neuroscience*, 2012, 7(3): pp. 274–281.

119 Nickerson, R.S., 'Confirmation bias: a ubiquitous phenomenon in many guises', *Review of General Psychology*, 1998, 2(2): pp. 175–220.

120 Bolsen, T., J.N. Druckman, and F.L. Cook, 'The influence of partisan motivated reasoning on public opinion', *Political Behavior*, 2014, 36(2): pp. 235–262.

121 Nestler, S., 'Belief perseverance', *Social Psychology*, 2010, 41(1): pp. 35–41.

122 Brehm and Cohen, *Explorations in Cognitive Dissonance*.

123 Martel, C., G. Pennycook, and D.G. Rand, 'Reliance on emotion promotes belief in fake news', *Cognitive Research: Principles and Implications*, 2020, 5(1): pp. 1–20.

124 Brady, W.J., et al., 'How social learning amplifies moral outrage expression in online social networks', *Science Advances*, 2021, 7(33).

125 Holman, E.A., D.R. Garfin, and R.C. Silver, 'Media's role in broadcasting acute stress following the Boston Marathon bombings', *Proceedings of the National Academy of Sciences*, 2014, 111(1): pp. 93–98.

126 Paravati, E., et al., 'More than just a tweet: the unconscious impact of forming parasocial relationships through social media', *Psychology of Consciousness: Theory, Research, and Practice*, 2020, 7(4): p. 388.

127 Baum, J. and R. Abdel Rahman, 'Emotional news affects social judgments independent of perceived media credibility', *Social Cognitive and Affective Neuroscience*, 2021, 16(3): pp. 280–291.

128 Clore, G.L., 'Psychology and the rationality of emotion', *Modern Theology*, 2011, 27(2): pp. 325–338.

129 Sulianti, A., et al., 'Can emotional intelligence restrain excess celebrity worship in bio-psychological perspective?', in *IOP Conference Series: Materials Science and Engineering* (IOP Publishing, 2018).

Index

Page numbers followed by (fn) indicate a footnote, e.g. 149(fn). Abbreviation: DB = Dean Burnett.

170–1, 175, 185–9 *see also* emotional contagion; empathy; facial expressions and emotions

communication technologies *see* phone calls; social media and online communication; therapeutic applications of technologies; video calls

confirmation bias 321, 324, 325

conformity 210, 219, 318, 325

consciousness, evolution of 82

consolidation of memories 96–8, 99, 100, 130–3

conspiracy theories 308–9, 316, 321, 328, 330 *see also* deception; misinformation and 'fake news'

constructed emotions theory 26, 31–3

corpus callosum (brain region) 34–5

cortex/neocortex (brain region) (in general) 36 *see also specific regions of the cortex*

cortisol 97

cross-race effect 183

crushes, in adolescence 258–9, 262

crying: DB's (in)ability to cry 13–14, 23–4, 33, 62–3, 206–7, 285; gender differences 207; induced by TV and films 62–3; types and functions of tears 14–15

cuteness and cute aggression 201–3

cyberbullying 275–6

dancing 127

Darwin, Charles 10, 27

deception: automated voices and announcements 299–300; response to 278, 299–300; self-deception 277–8 *see also* manipulation of emotions; misinformation and 'fake news'

defining 'emotions' 6–7, 9–13, 39, 84

deindividuation and 'mob mentality' 159–62

depression: caused by work 170, 171; gender differences 224–5; and gut microbiome 19; and memory 103; post-natal depression 194, 199; vagus nerve stimulation treatment 21

Diana, Princess of Wales, impact of death 246

digestive system, influence of 19, 21

disgust: as 'basic' emotion 28, 29; brain region associated with 148; and colour green 59; facial expression of 28, 29; and horror 26; and memory 105; and suppressed motivation 46 *see also* negative emotions

doctors, emotional aspects of work 186–9

dopamine 233

drama therapy 174–5

dreams and nightmares: AND model 137–8; bizarre nature of 131, 133; DB's bad dreams 129, 135, 139–40; due to COVID-19 pandemic 129, 135; and emotion processing 134–40; Freud's interpretations 133–4; and memory consolidation 130–3; and mental health 136; post-traumatic 136; prevalence of nightmares 129; recurring 138; threat simulation theory 135–6

Dunbar's number (of social relationships) 246–7

dysgranular field (brain region) 148–9

dysphoria 103 *see also* depression

e-learning, motivation in 51

earworms 123

Ekman, Paul 27–30, 31, 32, 296–7

Eleri, Carys 171–5

embarrassment 23, 29, 59, 79, 92, 97, 271 *see also* negative emotions

emojis and emoticons 304

emotion-cognition relationship: appraisal theory 164–6, 239–41; in attention and focus 69–70, 312–13, 314–15, 316–17; belief perseverance 325; cognitive dissonance 325; competition for brain's resources 160–2, 325; confirmation bias 321, 324, 325; distinction recognised by Stoics 7; in effect of emotions experienced 73–4; in empathy 178–81; in 'flow' state 272–4; interrelatedness (in general) 51, 74, 81–4, 329; in learning and information processing 51, 312–15, 316–17; in love 234–345; motivated reasoning 325; in motivation 44–7, 49, 77–8; negativity bias 316–17, 326; role of imagination 252–3; shared evolutionary origin 82; in stage fright 78–81

emotion-memory relationship: appraisal theory 164–6, 239–41; emotions triggered by memories 95–6, 99–100, 113, 134; fading affect bias 102–3, 124, 235; happy memories being more detailed 94; for implicit memories 92; later emotions changing memories 97–8, 102, 331; longevity and potency of emotional memories 90, 91, 92–4, 95, 102–3, 105; in memory consolidation 96–8, 100–1; in PTSD 72, 136, 138, 174, 306; role of nightmares 137–8; suppressing emotional memories 100–1

emotional contagion: dangers of 'mob mentality' 159–62; versus empathy 155–7; evolutionary importance 158–9; from groups of people 154–9, 161–2, 188; from music 117, 118, 157; neurological mechanism for 157; from social media 292–3

emotional detachment/suppression at work 163–4, 166–8, 170–1, 185–9

emotional manipulation *see* manipulation of emotions

emotional processing 70–3, 134–40

emotional regulation, brain regions responsible for 217, 218

emotional relationships: attachment during early childhood 80–1, 204; friendships 230, 246–7; one-sided *see* parasocial (one-sided) relationships; parent-baby emotional bond 193–5, 198–200, 203; role of neurotransmitters 195–200, 202–5; romantic *see* romantic relationships; *see also* social relationships

emotions: causing change 332–3; as conscious/subconscious processes 37–8; historical study of 7–11; identifying and defining 6–7, 9–13, 39, 84; language of 16, 25–6 *see also* categories and types of emotion; communicating and sharing emotions; emotion-cognition relationship; emotion-memory relationship; negative emotions; physiology-emotion connection; positive emotions; *specific emotions*

empathy: and autism 182, 214; in babies 149; and body language mimicry 150; versus emotional contagion 155–7;

evolutionary importance 145, 152–3, 156; influence of own emotions on 177–9; as ingrained 149, 152; ingroup versus outgroup bias 183–4; versus mentalising (theory of mind) 181; neurological mechanism for 118, 148–9, 156, 178–80; and physical pain 151–2; in romantic relationships 241–2; as selfish/unselfish 152–4

endocannabinoids 65

endocrine system 17–18 *see also* hormones

endorphins 15, 65

envy 58, 196 *see also* negative emotions

episodic memories 92, 122, 131

evaluative conditioning 121–2

excitation transfer theory 67

executive control 47, 81, 83 *see also* cognition (thinking)

existential dread, as a motivator 52–5

explicit memories 92

extrinsic versus intrinsic motivation 50–1

Facebook: DB's use of 267, 284–5; research into emotional manipulation 292, 294, 295 *see also* social media and online communication

faces, seeing in inanimate objects 301

facial colour changes 59–60

facial expressions and emotions: in artificial/CGI faces (uncanny valley) 302–3; automated emotion recognition 296–9; in cartoon characters (blinking) 301–2; cross-cultural similarities and differences 28–9, 30; difficulties distinguishing between emotions without context 30–1, 296, 297–9; early writings on 10; Ekman's work 27–30, 31, 32, 296–7; 'invisible' emotions 29; involuntary nature of expressions 26–7, 29, 30, 289; online curation of emotions portrayed 289–90

facial paralysis, and empathy 152

facial recognition, cross-race effect 183

facial recognition technology 296–9

fading affect bias 102–3, 124, 235

'fake news' *see* misinformation and 'fake news'

fandom 248, 259–62 *see also* parasocial (one-sided) relationships fear: as 'basic' emotion 28, 29; brain region associated with 39; enjoyment of 63, 67–8; facial expression and colour 28, 29; as first emotion 111; of flying 95–6; and horror 26; and imagination 253; and motivation 46, 49, 52–5; in PTSD 72; smell of (in sweat) 111 *see also* negative emotions

films and TV causing negative emotions 62–3, 67, 73, 74

Firth-Godbehere, Richard 6, 7, 10–11, 26, 197

flat Earth conspiracy theory 316, 321

'flow' state 272–4

flying, fear of 95–6

football shirts, red colour's competitive advantage 61

Freud, Sigmund 44, 133–4

friendships 230, 246–7 *see also* emotional relationships; social relationships

frosty atmospheres, emotional contagion 155, 157

funerals: crying at 206–7; of DB's father 155, 206–7, 245–6, 266–7, 284–5; emotional contagion at 155, 158; live streaming 266–7, 284–5

gender differences: adolescent crushes 258(fn); attitudes towards infidelity 239; in brains (beliefs and experimental studies) 209–26; in brains (DB's impossible experiment) 222–3; in emotional regulation and expression 207, 215, 217, 218, 222, 224; 'maternal instinct' 217; mental health problems 224–5; other physiological differences 207–9; societal influences 221–6

gender discrimination 211–13, 215, 224

goal distraction 61

green colour, associations and effects 58, 59, 61

grief: DB's acceptance of emotions 333–4; DB's anger 176–7, 191–2, 308; DB's attempts to disguise grief 142–4; DB's emotional confusion 2–3, 141–2, 206; DB's (in)ability to cry 13–14, 23–4, 33, 62–3, 206–7, 285; DB's motivation and productivity 42–3, 52; DB's need to talk after funeral 285–6; at death of Princess Diana 246; emotional processing 71, 333; shared grieving 141; stages of 177 *see also* negative emotions

guilt 29 *see also* negative emotions

habituation 236, 237

'hangry' behaviour 18–19

happiness 28

hippocampus (brain region): and dreaming 133; and emotional regulation 217; and emotions processing 38, 83, 231; and imagination 251–2; and memory 92–3, 95, 109–10, 133, 249; and navigation 107–8

Holmes and Rahe stress scale 227, 244

Holmes, Sherlock (analogy for action representation) 145–6

hormones: cortisol 97; digestive 19, 20; effect of tears on 15; influence on the brain (and emotions) 20; oestrogen 208, 218; oxytocin 15, 195–200, 202–3, 204, 205, 218; testosterone 208, 218, 219–21; vasopressin 197, 202–4, 218 *see also* endocrine system

horror (emotion) 26

horror movies 67, 73, 253

hypothalamus (brain region) 17, 45–7, 195

hysteria 212

Icke, David 319

identification, in parasocial relationships 255–6

identifying and defining 'emotions' 6–7, 9–13, 39, 84

imaginary friends 257–8

imagination and mental imagery 250–3, 257

imitation of observed actions 146–9

implicit memories 91

impression management 269–70, 277–80, 282, 289–91, 293–4

infancy and childhood: attachment with primary caregiver 80–1, 204; breastfeeding 198, 199; DB's memories of 104–5, 127–8; emotional experiences 82–3; empathy in babies 149; imaginary friends 257–8; importance of sense of smell 108–9; learning from media characters 256; nightmare frequency

136, 138; oxytocin in newborns 199; parent-baby emotional bond 193–5, 198–200, 203 *see also* adolescence and early adulthood

inferior frontal cortex (brain region) 147, 148

inferior parietal cortex (brain region) 118

infidelity, emotional versus sexual 239

insular cortex (insula) (brain region) 38, 148–9, 217, 232

intelligence, and brain anatomy 34–5

intention processing 146

intrinsic versus extrinsic motivation 50–1

intrusive thoughts 68–9

Izard, Carroll E. 11–12

jealousy 238

Kübler-Ross, Elizabeth 177

language of emotions 16, 25–6

language processing 118–19

learning (of information): from media characters 256; motivation 51, 309; from other people 310–12, 314–28; processing demands and information prioritisation 312–15, 316–17; from senses 310

LeDoux, Joseph 6

left brain/right brain facts and myths 34–6, 79, 147(fn), 178(fn)

limbic system (brain region) 36–9, 46, 107, 121

lobotomies 212–13

Lomas, Tim 25–6

London taxi drivers, brain study 107

losing oneself in a book/film 254–5

love: brain regions associated with 232, 233–5; demands on the brain 229; effect on cognition 234–5; for family and friends 230, 235(fn); role of dopamine 233; romantic love 229–38, 243, 244–5 *see also* romantic relationships

lust and sexual attraction: asexuality 231–2; brain regions associated with 231, 232; and romantic relationships 230–3, 237–8; Stoics' rejection of 7–8, 9; suppression of 48

Mack, Katherine (@AstroKatie) 53–5

mammal brain (region) 36–9, 46, 107, 121

manipulation of emotions: by authorities 294–5; for marketing 51, 255–6, 294, 295, 322; response to 278–9; by social media 292–4; by traditional news and media 314–15, 317, 327, 329

marketing 51, 255–6, 294, 295, 322

McCosh, James 11

medical work, emotional aspects 186–9

memory(ies): brain regions associated with 39, 91–2, 93–5, 250, 251; changeable nature of 32, 89, 97–8, 100, 101–2, 250–1, 331; connections with objects 99–100, 113, 265, 331; consolidation 96–8, 99, 100, 130–3; DB's memories of early childhood 104–5, 127–8; DB's memories of his father 89, 97–8, 99–100, 103–4, 113, 331; episodic memories 92, 122, 131; explicit memories 92; fading affect bias 102–3, 124, 235; forgetting memories 98–9; and imagination 251; implicit memories 91; and music 120–3; procedural memories 91–2; reminiscence bump 124; retroactive memory enhancement 98; semantic memories 92; and sleep 130–3; and smell(s) 104–5, 107, 108–12, 132; suppression of 100–1; as synapses 92–3, 99; working memory 313; Zeigarnik effect 90–1 *see also* emotion-memory relationship

mental health/illness: and social media 280–3; and status 280–1; therapeutic applications of technologies 305–6 *see also* anxiety; depression; PTSD; schizophrenia

mental imagery and imagination 250–3, 257

mentalising (theory of mind) 181, 234

mirror neurons 117–18, 146–7, 148, 157

mirroring body language 150

misinformation and 'fake news': about COVID-19 pandemic 308–9, 330; David Icke's space lizards 319; flat Earth theory 316, 321; and social media/internet 308–9, 315–16, 320–4, 326–8; susceptibility to 326 *see also* deception

'mob mentality' (deindividuation) 159–62
Moebius syndrome (facial paralysis) 152
monkey experiments, mirror neurons 117
Morgan, Matt 186–90
motivated reasoning 325
motivation: approach-attachment
behaviour 233–4; approach versus avoid
motivation 48–9, 51; brain regions
associated with 45–7, 48; and cognition
44, 45, 46, 47–8, 325; DB's experiences
during grief 42–3, 52; and emotions
42–7, 49, 51–5, 76, 86, 233–4; intrinsic
versus extrinsic motivation 50–1; and
novelty 68, 124, 309, 316
motivational salience 45
music: dancing 127; DB's emotional
response to 114, 121, 122, 127–8;
differentiating between voice and
instruments 126; earworms 123;
emotional contagion from 117, 118,
157; emotional response to 114–17,
118–27; evolutionary significance
125–6, 127; and memory 120–3
musical expectancy 118–19, 120

navigation, role of hippocampus 107–8
negative emotions: and attention/
focus 69–70, 79, 94, 316–17, 326;
and creativity 70; emotion processing
70–3, 134–40; feeling good whilst
experiencing 63, 67–8, 73; induced by
TV and films 62–3, 67, 73, 74; and
intrusive thoughts 68–9; and memory
94, 102–3, 105; as more impactful
than positive emotions 102; negativity
bias 316–17, 326; and novelty 68; and
performance 79 *see also specific emotions*
negativity bias 316–17, 326
nervous systems: enteric ('second brain')
19; parasympathetic 17–18, 20–1,
202–3; regulation by brain 17–18;
somatic and autonomic 17; sympathetic
17, 45, 202–3
neurotransmitters 15, 65, 195–200, 202–4,
205, 218
news and media (traditional): credibility
311–12, 318–20, 323, 329; emotional
content 314–15, 317, 327, 329;
precursors to 310–11 *see also* conspiracy

theories; misinformation and 'fake
news'; social media and online
communication
nightmares *see* dreams and nightmares
noises, emotional response to 115, 116,
125–6 *see also* music
novelty 68, 124, 268–9, 309, 316

objects, and memories 99–100, 113, 265,
331
oestrogen 208, 218
olfactory bulb and cortex (brain region)
106, 107, 109–10
olfactory system 106, 107, 108, 109–12
one-sided relationships *see* parasocial
(one-sided) relationships
online communication *see* social media
and online communication
online learning, motivation in 51
orbitofrontal cortex (brain region) 38, 48
oxytocin 15, 195–200, 202–3, 204, 205,
218

pain (physical): and empathy 151–2;
enjoyment of 64–6, 67
paracingulate sulcus (brain region) 156
parasocial (one-sided) relationships:
adolescent crushes 258–9, 262; benefits
256–9, 263; ending the relationship
262–3; with fictional characters 248,
253–5, 259–62, 301–2; identification
with the object 255–6; with imaginary
friends 257–8; losing oneself in a
narrative 254–5; meeting the object
259–62; negative aspects 259–62;
neurological mechanisms 253–4; with
people you haven't met 246, 248–9,
253–4, 255–6, 295
parasympathetic nervous system 17–18,
20–1, 202–3
Parch (TV drama), actor's experiences
171–4
pareidolia 301
parent-baby emotional bond 193–5,
198–200, 203 *see also* attachment during
early childhood
'passions' 7–8, 9, 10
pathos 8
performance anxiety (stage fright) 78–81

283–4; emojis and emoticons 304; and emotional contagion 292–3; impression management 269–70, 277–80, 282, 289–91, 293–4; lack of nonverbal emotional cues 286, 288; live streaming funerals 266–7, 284–5; machine detection of emotions 304; negative aspects 274–82, 308–9, 315–16, 320–4, 326–8; online versus in-person emotions and personae 290–1; positive aspects 269–74, 282–3; versus real-world interactions, cognitive demands 270–1; and reward 268–9, 271, 324; and self-deception 277–8; and self-validation 267–8, 272, 324; and status 269–70, 280–1 *see also* conspiracy theories; Facebook; misinformation and 'fake news'; news and media (traditional); video calls

social relationships: cognitive load associated with 247, 270–1; Dunbar's number 246–7; friendships 230, 246–7; one-sided *see* parasocial (one-sided) relationships; *see also* emotional relationships

somatic marker hypothesis 22–3

spicy food, enjoyment of pain caused by 64

spindle cells 156

Spiner, Brent 259–60

sports kit, competitive advantage of wearing red 61

SPOT (Screening of Passengers by Observation Techniques) programme 296–9

stage fright 78–81

stalkers 261–2

Star Trek: The Next Generation (TV series): Data actor's experience of fans 259–60; Data's inability to choose ice-cream flavour 85–6

Star Trek (TV series): Stoicism of Vulcans 8, 41; universal use of English language 28

Starbucks (branding) 51

status: and emotions 66; and social media 269–70, 280–1; subjective status and mental health 280–1

Stoics and Stoicism 7–9, 41

stress: benefits of green environments for 61; caused by uncertainty 52; caused by work 168, 186; coping mechanisms 166; cortisol 97; Holmes and Rahe stress scale 227, 244; PTSD 72, 136, 138, 174, 306; and status 61; Yerkes-Dodson curve 79 *see also* anxiety; negative emotions

striatum (brain region) 91–2

study of emotions (historical) 7–11

suicide 188, 224–5

superior temporal cortex (brain region) 146, 147, 148, 181

suppression of emotions: during disagreements with romantic partner 239–41; in learning and decision making (as impossible) 328–9; at work 163–4, 166–8, 170–1, 185–9

supramarginal gyrus (brain region) 178–9

surprise 28, 29

sympathetic nervous system 17, 45, 202–3

synapses (neuron connections) 92–3, 99

taxi drivers, brain study 107

tears, types and functions of 14–15 *see also* crying

teenage years *see* adolescence and early adulthood

temporal lobe (brain region) 46, 252

testosterone 208, 218, 219–21

thalamus (brain region) 109, 231, 249

theories of emotions *see* basic emotions theory; constructed emotions theory

theory of mind (mentalising) 181, 234

thinking *see* cognition (thinking)

threat simulation theory 135–6

transportation phenomenon 254–5

triune brain model 36, 83

TV and films causing negative emotions 62–3, 67, 73, 74

types of emotion *see* categories and types of emotion

uncanny valley 302–3

uncertainty, unpleasant nature of 52

vagus nerve 20–1

valence (component of affect) 45, 63

vasopressin 197, 202–4, 218